principles

of

electronic communications

PRENTICE-HALL SERIES IN ELECTRONIC TECHNOLOGY

Dr. Irving L. Kosow, *editor*

Charles M. Thomson, Joseph J. Gershon, and Joseph A. Labok,
consulting editors

principles
of
electronic communications

MATTHEW MANDL

Prentice-Hall, Inc., Englewood Cliffs, N.J.

Library of Congress Cataloging in Publication Data

MANDL, MATTHEW.
    Principles of electronic communications.

    (Prentice-Hall series in electronic technology)
    1. Telecommunication. 2. Signal theory
(Telecommunication) 3. Electronics. I. Title
II. Title: Electronic communications.
TK5101.M28   621.38   72-5573
ISBN 0-13-709030-7

© 1973 by Prentice-Hall, Inc.
Englewood Cliffs, New Jersey

10 9 8 7 6 5 4 3 2 1

Printed in the United States of America

PRENTICE-HALL INTERNATIONAL, INC., *London*
PRENTICE-HALL OF AUSTRALIA, PTY. LTD., *Sydney*
PRENTICE-HALL OF JAPAN, INC., *Tokyo*
PRENTICE-HALL OF CANADA, LTD., *Toronto*
PRENTICE-HALL OF INDIA PRIVATE LTD., *New Delhi*

contents

preface

Principles of Electronic Communications is a text designed to provide broad coverage in major communication areas, both from the standpoint of the basic circuitry involved and the unification of the components which make up complete systems. As such, the discussions are confined primarily to those topics that are closely related to communications. Consequently, course material on basic electronics has been omitted, and it is presumed that the user will have an adequate grounding in electronic fundamentals plus fundamental circuit theory.

Where fundamental concepts are given, or basic mathematical approaches detailed, the purpose is clarification of circuit characteristics to enable the reader to acquire a clearer understanding of how these factors relate to communications and how they aid in making up the complete transmitting or receiving system.

The bulk of the explanatory mathematics is contained in the initial chapters to provide a clear concept of signal synthesis, harmonic relationships, and the methods employed for utilizing signals of specific characteristics to obtain desired levels and types of modulation.

Specific circuitry found in communications and operational theory is covered after Chapter 3, with AM, FM, and TV systems and associated components included in Chapters 6 through 9. For the FM coverage in Chapter 7, stereo multiplexing is also discussed, including SCA methods and multiplexing, both from the standpoint of transmitting and of receiving.

Chapters 10 through 12 cover filters, transmission lines, microwave principles, and antenna systems, to round out the complete communication field. Practical review questions are included at the end of each chapter, with most chapters also including problem-type questions to indicate theory and mathematical applications in field work.

The Appendices include a summary of telemetry principles, plus discussions on pulse-modulation methods. Other sections contain mathematics tables, radio and TV frequency allocations, and other pertinent data.

Matthew Mandl

1
signals and spectra

1-1. introduction

Throughout the ages man has distinguished himself from the lower
forms of animal life by his sophisticated verbal and written communication
skills. Early in his history he relied heavily on vocal abilities and gestures to
convey his thoughts to those near him. As the complexity of mankind's society grew, however, the need for communication over greater distances became
more imperative. Early attempts to improve sound coverage (besides shouting)
included the beating of drums or the use of horns and megaphones. To increase the range, visual means were employed, such as smoke signals, flag
waving, and (in darkness) the waving of lighted flares.

The inherent limits of such methods, however, had to be overcome as
society flourished and required communications over yet greater distances.
As literacy became more prevalent, the carrying of written messages spread
and eventually evolved into the global mail system we have today. Discoveries
in the electric–electronic fields and the gradual increase in our knowledge
and utilization of such principles finally permitted direct and immediate
communication systems for both sound and sight.

In man's slow and arduous progress to the present-day marvels of his
various electronic communication systems, he had to explore and discover,
in a virtual step-by-step fashion, the potentials for communication that existed in the spectra, besides those of sound and light with which he was familiar.
Initially, he had to analyze and understand more thoroughly the characteristics of our own abilities to distinguish sound and light waves. He soon dis-
covered a curious phenomenon of nature, that our receptive senses of sound
and sight are sensitive to wave frequencies separated by an incredibly
enormous span.

When we speak, make a noise, or play a musical instrument, we generate variations in air pressures, which we call sound waves. When these reach the human ear, its drum is vibrated and the auditory nerve transmits a signal to the brain, enabling sound recognition. The frequency range of such sounds lies at the bottom of the vast frequency spectrum known to us at present, and represents only a fractional segment.

Our eyes, on the other hand, enable us to convert certain waves to sensations of light. Such waves, however, are removed from the sound waves by a frequency span of more than 8×10^8 megahertz (MHz).*

Another of our senses, touch, enables us to recognize an additional sector of the frequency spectrum, that of heat, which lies below the visible spectrum, but is still separated from sound by a great frequency span. Other than these three portions of the vast spectrum, however, our inherently endowed facilities fail us. Direct utilization of sound frequencies limits communication, but light frequencies do not (we can see stars many light years distant). Thus, between the sound and light frequencies we would expect to find portions of the spectrum with varying communication potentials in terms of usable distances.

This, indeed, has been the case, and man, by his ingenuity, has utilized his accumulated scientific knowledge to devise numerous electronic communication systems operating in various segments of the spectra so that he can transmit and receive data over virtually unlimited distances. Consequently, his achievements in visual as well as audible signal transmissions have become a vast communications network in our modern society and an integral part of our daily lives in the realms of direct personal dialog between individuals, the dispersion of news, scientific and commercial data retrieval, and entertainment. Such communication systems, because of their convenience and frequency of use, have become such an accepted routine in our daily lives that the average participant gives little thought to their almost miraculous properties.

In the study of various communication systems it is important to relate the operation of the individual units that make up a system to that portion of the frequency spectrum in which it operates. Hence, to obtain such proper perspective, this chapter covers the fundamental aspects of electromagnetic waves and spectra. Also, since signal waveforms are involved with virtually all circuitry, it is essential that the fundamental characteristics of sinewaves, complex waves, sawtooth and square waves, and other signals be understood. Similarly, the relationship of harmonics in signal content to signal makeup should be known. Only then can progressive circuit analysis become more meaningful.

Hertz (Hz) is used throughout this text instead of *cycles per second* (*cps*) to conform to the universal standards for such designations.

In subsequent chapters certain areas of these discussions will be covered in greater depth as they apply directly to specific circuitry or component packages making up the complete transmitting or receiving gear.

1-2. electromagnetic waves

Sound waves represent variations in air pressure above and below an average level at a rate determined by the frequency (pitch) of the sound. The degree of change is determined by the loudness of the sound. "Radio" waves used for the transmission and reception of electronic signals are composed of electrostatic and electromagnetic fields that make up a propagated wavefront the amplitude of which relates to the energy content and the rate of change to its fundamental frequency. The characteristics of such waves are discussed more fully in Chapter 12.

A sound or radio wave with a specific frequency is represented by a sinewave, as shown in Fig. 1-1, where a positive and negative alternation

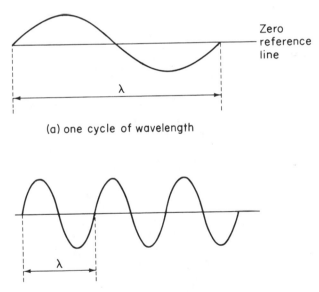

(a) one cycle of wavelength

(b) increased frequency

figure 1-1 wavelength and frequency

make up a complete cycle. When such waves are propagated, they span distances in a time interval dependent on velocity factors. *Velocity* is the time rate of motional change of position in a specific direction. Velocity is often used loosely as a synonym for *speed*, though actually speed is the time rate of change of position in a given direction.

The velocity of sound depends on the medium through which it travels (air, water, etc.) and on temperature. In air an approximation for the velocity of sound is 1100 feet per second (ft/s), or 750 miles per hour (mph). At 32° F the velocity is 1088 ft/s. In the metric system the velocity at 0°C is 349.8 meters per second (m/s).

The velocity for light is identical to that for electromagnetic radio waves and is 299,792.5 kilometers per second (km/s), or 186,282 miles/s in vacuum.

Wavelength is the distance traveled by one cycle of a waveform in a given period. Thus, if a sound wave has a velocity of 331 m/s, and one cycle is produced per second, the wavelength is 331 m. Thus, for the single cycle shown in Fig. 1-1a, the wavelength (λ) is the span between the beginning and end of the cycle and represents a distance of 331 m. As the frequency (Hz) is increased, wavelengths become shorter, as shown in Fig. 1-1b. For a frequency of 3 Hz, the wavelength would be one third of 331, or 110 m. Wavelength (λ), velocity (v), and frequency (f) are related as shown by the following equations:

$$v = \lambda f \tag{1-1}$$

$$f = \frac{v}{\lambda} \tag{1-2}$$

$$\lambda = \frac{v}{f} \tag{1-3}$$

When setting up formulas for light waves and radio waves, the velocity expressed in meters is usually rounded off to 300,000,000 m/s, and the velocity in miles to 186,000 miles/s. Based on Eqs. (1-1) to (1-3), we obtain

$$\lambda = \frac{300,000,000}{f(\text{Hz})} = \text{m} \tag{1-4}$$

$$\lambda = \frac{186,000}{f(\text{Hz})} = \text{miles} \tag{1-5}$$

$$\lambda = \frac{984}{f(\text{MHz})} = \text{ft} \tag{1-6}$$

$$\lambda = \frac{11,808}{f(\text{MHz})} = \text{in.} \tag{1-7}$$

$$\lambda = \frac{300,000}{f(\text{kHz})} = \text{m} \tag{1-8}$$

$$f = \frac{300,000}{\lambda(\text{m})} = \text{kHz} \tag{1-9}$$

$$\lambda = \frac{300}{f(\text{MHz})} = \text{m} \tag{1-10}$$

$$f = \frac{300}{\lambda(\text{m})} \qquad = \text{MHz} \qquad (1\text{-}11)$$

Equation (1-6) is obtained by multiplying 186,282 by the number of feet in a mile (5280) to obtain 983,568,960 and pointing off six places so that frequency in megahertz can be used. Similarly, 984 is multiplied by 12 to obtain 11,808 in Eq. (1-7) to provide for the answer in inches. Such formulas, for short wavelengths, are useful in the higher-frequency portions of the spectrum for calculating transmission line, waveguide, and antenna lengths, as illustrated in detail in Chapters 10, 11, and 12.

Conversion factors can be used to change meters to feet if desired, instead of employing specific equations. If, for instance, we use Eq. (1-10) to solve for wavelength when the frequency is 6 MHz, we obtain

$$\lambda = \frac{300}{6\,\text{MHz}} = 50\,\text{m}$$

Now, if we also require the wavelength in feet, we could use Eq. (1-6):

$$\lambda = \frac{984}{6\,\text{MHz}} = 164\,\text{ft}$$

After having obtained 50 m from Eq. (1-10), however, the multiplication of meters by 3.2808 ft/m would have provided the wavelength in feet:

$$50\,\text{m} \times 3.2808\,\text{ft/m} = 164.04\,\text{ft}$$

Similarly, the multiplication of the number of feet by 0.3048 m/ft produces the wavelength in meters:

$$164\,\text{ft} \times 0.3048\,\text{m/ft} = 49.98\,(50\,\text{m})$$

For centimeters (cm), the number of feet is multiplied by 30.48 cm/ft:

$$164\,\text{ft} \times 30.48\,\text{cm/ft} = 4998.72\,\text{cm}$$

Table 1-1 includes the foregoing and other conversion factors for convenience in changing the units of one system to that of the other. The results obtained from such conversion tables are close approximations because of round-off factors (sufficiently close to the proper answer for most practical purposes). Note, for instance, that the multiplier for converting miles to meters is 1609.3 m/mile. However, if we use the *feet to meters* multiplier 0.3048 and multiply it by 5280, we arrive at a more accurate multiplier of 1609.344.

Note that the micron is one millionth of a meter, while the millimicron is one-thousandth part of a micron. The angstrom unit is also known as the *tenth-meter* (10^{-10} meter) and its symbol is Å. Angstrom units are used to designate wavelengths in meters or fractional meters in spectrum charts in addition to the frequency designations, as shown in Section 1.3.

table 1-1   conversion factors for length

mult. no. of	by	to obtain no. of
inches	2.540	centimeters
inches	0.02540	meters
feet	30.48	centimeters
feet	0.3048	meters
miles	5280.0	feet
miles	1.6093	kilometers
miles	1609.3	meters
centimeters	0.3937	inches
centimeters	0.01	meters
centimeters	10.0	millimeters
meters	100.0	centimeters
meters	3.3808	feet
meters	39.37	inches
meters	1000.0	kilometers
microns	10^{-6}	meters
microns	10^{-4}	centimeters
millimicrons	10^{-7}	centimeters
angstroms	10^{-10}	meters
angstroms	10^{-8}	centimeters

Also of value are velocity conversion factors relating to seconds, minutes, and hours, as shown in Table 1-2.

table 1-2   conversion factors for velocity

mult. no. of	by	to obtain no. of
feet/second	1.097	kilometers/hour
feet/second	0.6818	miles/hour
feet/second	0.01136	miles/minute
centimeters/second	1.969	feet/minute
centimeters/second	0.036	kilometers/hour
centimeters/second	0.02237	miles/hour
miles/hour	44.70	centimeters/second
miles/hour	88.0	feet/minute
miles/hour	1.467	feet/second
miles/hour	26.82	meters/minute

1-3. spectra

The audible-frequency spectrum is shown in Fig. 1-2, and ranges from approximately 15 Hz to 16 kilohertz (kHz). Below 15 Hz vibrations are felt

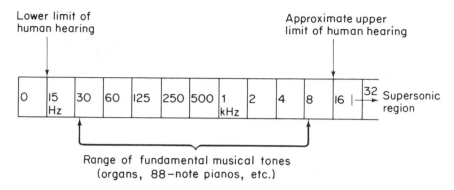

figure 1-2 audible spectrum

rather than heard, and above 10 kHz the degree of hearing depends on age. Overtones (harmonics) of musical sounds extend beyond the range of fundamental tones shown in Fig. 1-2 and are related to the fundamental tone by being even or odd multiples of the fundamental frequency. (Harmonics are discussed more fully later.)

The visible spectrum is shown in Fig. 1-3 and ranges from approximately

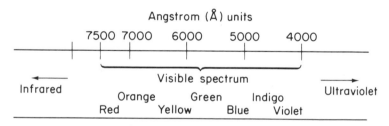

figure 1-3 visible spectrum

7500 to 4000 Å. The infrared region is at the low-frequency end of the visible spectrum, and the ultraviolet at the higher end. Electronic sensors of various types are capable of picking up such frequency signals beyond human sight.

The relative position of the visible-frequency region with respect to the inclusive spectrum is shown in Fig. 1-4. In the intervening frequencies between the audible and visible regions lie the various comunications systems authorized by the Federal Communications Commission (FCC); they include ship to shore, public-service broadcasts (police and fire), amateur, aircraft, entertainment (TV, FM, and radio), and others. Above the visible-frequency region, beginning in the shorter wavelengths of the ultraviolet, we enter portions of the spectrum that are dangerous to humans, depending on the degree and length of exposure.

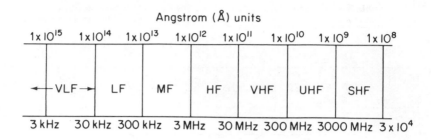

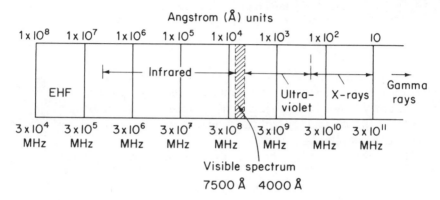

figure 1-4 inclusive frequency spectrum

Certain designations have been given to portions of the spectrum for convenience in referring to them (see Table 1-3).

table 1-3   frequency designations

VLF (very low frequencies)	3 Hz	to	30 kHz
LF (low frequencies)	30 kHz	to	300 kHz
MF (medium frequencies)	300 kHz	to	3 MHz
HF (high frequencies)	3 MHz	to	30 MHz
VHF (very high frequencies)	30 MHz	to	300 MHz
UHF (ultrahigh frequencies)	300 MHz	to	3000 MHz
SHF (superhigh frequencies)	3 GHz	to	30 GHz*
EHF (extra-high frequencies)	30 GHz	to	300 GHz

*See Appendix J.

Alternative frequency designations used by the military in the past are shown in Table 1-4.

table 1-4   military frequency designations

P band	225 MHz	to	390 MHz
L band	390 MHz	to	1550 MHz
S band	1550 MHz	to	5200 MHz
X band	5200 MHz	to	10,900 MHz
K band	10,900 MHz	to	36,000 MHz
Q band	36 GHz	to	46 GHz
V band	46 GHz	to	56 GHz

1-4.   signals and harmonics

A microphone is a *transducer* that converts sounds into electric signals. If we attempted to radiate such signals by amplifying them and feeding them to a large wire loop (to simulate an antenna), transmission would occur for only limited distances. Hence, the enormous power required to increase distance makes such a system impractical. The upper-frequency signals of the spectrum used for broadcast purposes, however, propagate at the speed of light and are capable of spanning great distances. They can be generated and processed with only a fraction of the power needed to send audio signals over short distances.

Thus, to transmit audito, video, or other low-frequency signals, it is necessary to use them to *modulate* the upper-frequency signals (known as *radio-frequency* signals). Thus, in essense, the radio-frequency (RF) signals "carry" the low-frequency information and hence are known as the *carrier*. The basic process is shown in Fig. 1-5.

A variety of methods can be used to alter a carrier signal in order to transmit required information. Included are amplitude modulation (AM), frequency modulation (FM), and phase modulation (PM). In any modulation process in which the characteristics of the carrier wave are altered, additional signals are produced. Such signals are called *sidebands* and are frequency related to the fundamental carrier signal, as discussed in Chapter 2.

In virtually all signal generating and modulating processes, other signals, known as *harmonics*, are also generated. Harmonics (which also bear a relationship to the sideband-production process) may be of a spurious nature and unwanted, or their presence may be desired for frequency-multiplication and other purposes, as discussed in subsequent chapters. Hence, the underlying characteristics and basic principles of harmonics should be understood so that the factors of their utilization or rejection can be associated with the study of oscillators, amplifiers, modulators, and other transmitting circuitry.

A pure signal (whether audio or RF) is one having only a single fre-

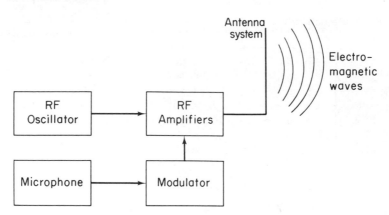

figure 1-5 basic transmitting system

quency with characteristics as shown in Fig. 1-6a. Such an alternating-current (ac) signal is composed of successive alternations (half-cycles) each of which has the same amplitude and width as the others and a gradual incline and decline in both the positive and negative directions. *Any deviation* from these specifications produces signal *distortion*, which, in turn, generates harmonics (signals or frequencies other than the fundamental).

A typical example of a waveform with a high harmonic content is shown in Fig. 1-6b. Amplitude differences in the alternations as well as varying widths generate frequencies in addition to the fundamental. In Fig. 1-6c distortion occurs in the decline of the positive and negative alternations and hence also indicates harmonic content. Square-wave and sawtooth signals, as shown in Figs. 1-6d and e, also have a high order of harmonic content. Both the steepness of the vertical components of the signals, and the *degree of abrupt change* in polarity, depend on the number of higher-frequency components present.

The numerous harmonics in musical tones provide the distinguishing features that permit recognition of individual musical instruments. A saxophone and violin, when playing the musical tone *A* at 440 Hz, each generate the same fundamental frequency but very different harmonics. Hence, each instrument produces a tone quality characteristically its own and recognizable from the other.

Harmonics are higher in frequency than the fundamental signal, but usually lower in amplitude. Harmonics have decreasing amplitudes for higher harmonic frequencies. How harmonics combine with the fundamental-frequency signal to produce a composite-frequency signal is shown in Fig. 1-7. The fundamental-frequency signal is shown in Fig. 1-7a and could be a very low frequency (VLF) audio signal or an ultrahigh RF signal.

The signal shown in Fig. 1-7b has a frequency three times that of the fundamental signal and thus represents a third harmonic. When these two

Each alternation has: Same width   Gradual
incline and decline

Same
amplitude

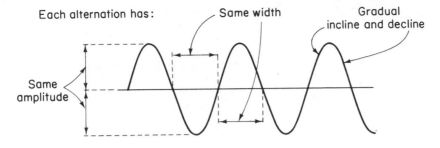

(a) pure sinusoidal waveform

(b) waveform with high harmonic content

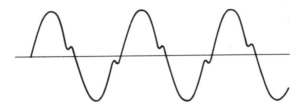

(c) distortion at decline of positive and negative alternations

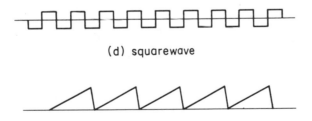

(d) squarewave

(e) sawtooth wave

figure 1-6 pure and distorted waveforms

signals are combined, the resultant is as shown in Fig. 1-7c. At any point in time when the signals are not in phase, the resultant will have a lower amplitude. Where in-phase conditions prevail, the amplitude of the signal would rise. Additional harmonic components would modify the resultant signal to a greater extent.

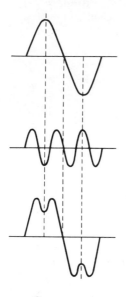

(a) fundamental (pure) frequency signal

(b) third harmonic

(c) sum of fundamental and third harmonic

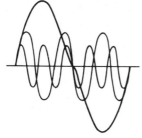

(d) sum of fundamental, third, and fifth harmonic

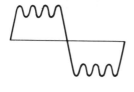

(e) sum of fundamental, third, fifth, and seventh harmonic

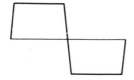

(f) squarewave synthesized

figure 1-7 signal synthesis

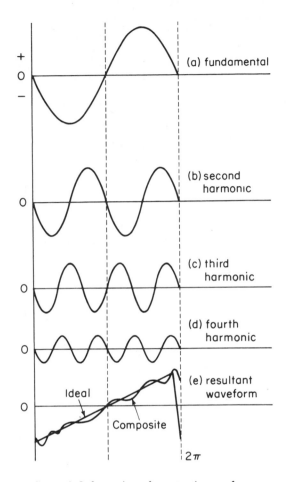

figure 1-8 formation of sawtooth waveform

In Fig. 1-7d, a fundamental signal is shown plus a third and fifth harmonic. Note that at the start of the combination waveforms, there are in-phase conditions for all three signals. Consequently, the composite signal has a steep rise time. When successive odd-harmonic signals are added to the fundamental, the composite signal starts to resemble a square wave.

Figure 1-7e shows a signal having a fundamental, third, fifth, and seventh harmonic content. With a sufficient number of higher-order odd-harmonic components, the resultant is a square wave and appears as shown in Fig. 1-7f.

Sawtooth signals of the type shown in Fig. 1-6e have *both* odd and even harmonic frequency signal components. This is illustrated in Fig. 1-8a, where the fundamental is shown. The addition of the second, third, and fourth harmonics produces a resultant as shown in Fig. 1-8e. With a greater number of higher harmonics, the incline of the sawtooth would become more nearly a straight line.

1-5. Fourier series

As mentioned earlier and illustrated in Fig. 1-6a, the sinewave signal is a single-frequency type; hence, its derivative (representative of the rate of change) is of the same form as itself (as is also the integral). With a signal of harmonic content it is possible to analyze the complex waveform by the calculus and obtain information on the particular harmonics making up the composite signal, thus ascertaining relative phase and amplitude factors. For analytical purposes the complex waveform may be an oscilloscope pattern or photograph, a graphed version produced by an X-Y plotter, tabulated values of fixed points on a waveform, or from a stated equation relating to the signal.

In the analysis of a complex waveform having periodic characteristics, the Fourier series is a convenient mathematical tool in design practices and in analytical surveys of systems. Commercial wave analyzers and distortion meters are widely used in industry for electronic waveform synthesis and measurement of total harmonic distortion. A knowledge of the Fourier series, however, helps us understand the nature of harmonics, complex waves, and factors relating to signal distortion. Since good groundwork can be obtained by employing the Fourier series with a stated equation, we shall confine our discussions to this method.

For a clearer understanding of Fourier analysis, a brief review of instantaneous value factors is helpful. As shown in Fig. 1-9a, instantaneous amplitudes of a sinewave can be related to a rotating phasor arm having a length (radius of circle) equal to the peak amplitude of the waveform. By trigonometric functions we can find the instantaneous value of the voltage *or* current by multiplying the maximum amplitude by the *sine* of the angle.

$$e = E \sin \theta \qquad (1\text{-}12)$$

where e = instantaneous value of voltage

E = maximum (peak) value of voltage

θ = angle formed by the vector arm and the zero voltage line

The same factors apply to instantaneous current (i).

$$i = I \sin \theta \qquad (1\text{-}13)$$

If, in mathematical analysis, we are interested primarily in relative amplitudes without particularly identifying them with voltage or current, we can restate the formula as

$$a = A \sin \theta \qquad (1\text{-}14)$$

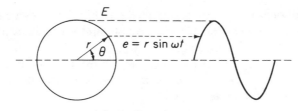

(a) rotating phasor

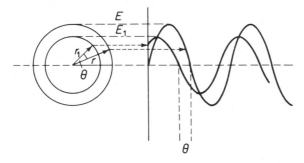

(b) two rotating phasors displaced by angle θ

$$e = r \sin \omega t \qquad e_1 = r_1 \sin (\omega t + \theta)$$

figure 1-9 instantaneous values of E or I

With respect to time (t) in seconds, we can also use $2\pi f$ (6.28f), which is known as the *angular velocity* (ω) because the phasor arm shows continuous changes of degrees with instantaneous amplitudes, and hence is indicative of the velocity of the sinewave signal:

$$e = E \sin \omega t \tag{1-15}$$

where e = instantaneous voltage value
E = maximum voltage value
ω = angular velocity $2\pi f$
t = time in seconds

Again, we can use i or a as desired in Eq. (1-15). If we are involved with fundamental as well as harmonic signals, we can identify peak amplitude, voltage, or current by using subscripts for identification:

$$a_1 = A_1 \sin \omega t \text{ (fundamental)}$$
$$a_2 = A_2 \sin \omega t \text{ (second harmonic)}$$
$$a_3 = A_3 \sin \omega t \text{ (third harmonic)}$$

Since the radius of the phasor arm equals peak amplitude, we can substitute r in the formula, as shown in Fig. 1-9a (for either amplitude, voltage, or current calculations):

$$a = r \sin \omega t \tag{1-16}$$

Phase relationships are shown in Fig. 1-9b. Assume that radius vectors r and r_1 have the same angular velocity, but are separated by a fixed angle θ, known as the *phase angle*. When $t = 0$, one waveform is at zero amplitude and starts to rise in a positive-amplitude direction and is represented by $e = r \sin \omega t$. The other waveform, however, leads this position by angle θ and hence is expressed as

$$e_1 = r_1 \sin(\omega t + \theta) \tag{1-17}$$

The evaluation of the signal components of a complex waveform involves both magnitude and phase. A signal component that reaches zero value 90° later than the instant when $t = 0$ is as shown in Fig. 1-10a, and

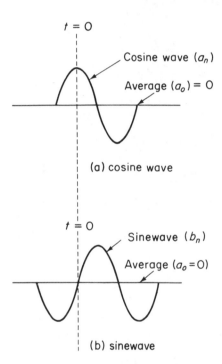

(a) cosine wave

(b) sinewave

figure 1-10 quadrature and in-phase components

is known as the *quadrature* component. If we substitute a for r to represent *peak amplitude*, this component can then be expressed as

$$a_n \cos \omega t \tag{1-18}$$

To distinguish the sine component (shown in Fig. 1-10b) from the quadrature, b is used to indicate peak amplitude:

$$b_n \sin \omega t \tag{1-19}$$

Thus, the leading or lagging aspects of a signal, plus amplitude factors, are evaluated by adding the two components that are in quadrature to obtain

$$r \sin (\omega t + \theta) = a_n \cos \omega t + b_n \sin \omega t \qquad (1\text{-}20)$$

The Fourier series represents a complex wave by an expression composed of *average* (a_o), *cosine* (a_n), and *sine* (b_n) terms:

$$f(t) = a_o + a_1 \cos \omega t + a_2 \cos 2\omega t + a_3 \cos 3\omega t \cdots + a_n \cos n\omega t \cdots$$
$$ + b_1 \sin \omega t + b_2 \sin 2\omega t + b_3 \sin 3\omega t \cdots + b_n \sin n\omega t \cdots \qquad (1\text{-}21)$$

The a_o designation is the dc term (average value). The a_1 and b_1 of the cosine and sine series represent the *fundamental* signal component of the complex waveform. Thus, as the lowest frequency signal contained in the complex waveform, it also represents the fundamental frequency of the complex signal itself and must be considered in any Fourier series. Higher subscripts for the a and b designations indicate higher-order harmonics of the fundamental (integer multiples of the fundamental signal frequency). The Fourier series can also be expressed as

$$\frac{a_o}{2} + \sum_{n=1} (a_n \cos n\omega t + b_n \sin n\omega t) \qquad (1\text{-}22)$$

The equations for solving values of a_o, a_n, and b_n for a one-cycle span of a complex waveform between the limits of $\omega t = 0$ and $\omega t = 2\pi$ are

$$a_o = \frac{1}{\pi} \int_0^{2\pi} f(t) \, d(\omega t) \qquad (1\text{-}23)$$

$$a_n = \frac{1}{\pi} \int_0^{2\pi} f(t) \cos n\omega t \, d(\omega t) \qquad (1\text{-}24)$$

$$b_n = \frac{1}{\pi} \int_0^{2\pi} f(t) \sin n\omega t \, d(\omega t) \qquad (1\text{-}25)$$

These are known as the *coefficients* of the Fourier series. Depending on the characteristics of a particular complex waveform, sine (or cosine) terms may have zero amplitude; hence, Eq. (1-24) or (1-25) will not be used. Also, all even (or all odd) harmonics may be absent, thus simplifying the analytical processes. Often these factors can be determined by inspecting the waveform and relating its configuration to any symmetrical characteristics that are evident.

Consider, for instance, the complex waveform shown in Fig. 1-11a. Here the negative portion of the signal (below the zero-amplitude horizontal line) has a different contour from the positive signal. If we take any horizontal span between the incline and decline of this waveform and draw a straight line, we have established two points for indicating symmetry. The horizontal line is divided into two equal parts by the vertical line marked f in Fig. 1-11a, and symmetry around the vertical axis is evident. Such *vertical-axis symmetry* is known as an *even* function, and the sine constants (b_n) are all zero. For a complex waveform with such axis symmetry, note that the point to the left of

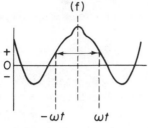

(a) complex waveform with
vertical axis symmetry

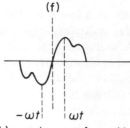

(b) complex waveform with
point symmetry

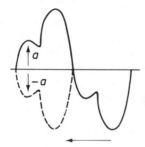

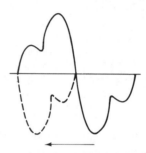

(c) waveform containing only
odd harmonics

(d) waveform containing even
harmonics

figure 1-11 axis and point symmetrical factors

the vertical axis ($-\omega t$) has the same *amplitude* and electric polarity at the point (ωt) to the right of the vertical (f) axis.

Since the sine constants are zero, the waveform contains only cosine values (a_n) and dc (a_o). Cosine terms have identical algebraic signs for the $-\omega t$ and ωt shown in Fig. 1-11a. Sine terms, on the other hand, would have opposite algebraic signs for the $-\omega t$ and ωt points.

For the type of waveform shown in Fig. 1-11b, the vertical line (f) divides the waveform equally between its positive and negative sections. Note that equidistant points from the vertical axis ($-\omega t$ and ωt) have the same amplitude but are of opposite electric polarity, as opposed to those of Fig. 1-11a, where the two points have the same polarity. Thus the waveform in Fig. 1-11b has *point symmetry* (symmetrical around any point of intersection on the horizontal axis) and is known as an *odd* function. A signal with these characteristics indicates that the cosine constants (a_n) are all zero, and only sine (b_n) terms and direct current are used in the Fourier series.

Often whether a complex wave contains only even (or odd) harmonics can also be determined by inspection of the characteristics of the signal. As mentioned earlier, a periodic complex wave combines signals having frequencies that are integral multiples of the fundamental signal frequency.

Periodicity is lost if we combine signals having frequencies that are not such integral multiples of the fundamental.

Waveforms containing only odd harmonic signals are symmetrical around the horizontal axis; that is, if we shift the negative-polarity portion a half-cycle to the left, as shown in Fig. 1-11, instantaneous amplitudes of each wave portion are identical except for polarity. Note that this is not the case for the signal shown in Fig. 1-11d, indicating even-harmonic components are present. (Point symmetry is also present in Fig. 1-11d, indicating b_n sine functions only, the same as in Fig. 1-11b.)

The reasons for horizontal-axis symmetry are illustrated in Fig. 1-12.

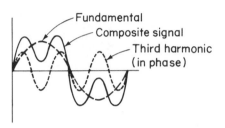

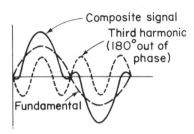

(a) third harmonic in phase (with fundamental)

(b) third harmonic 180° out of phase

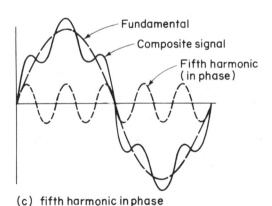

(c) fifth harmonic in phase

figure 1-12 symmetrical, odd-harmonic composite signals

In Fig. 1-12a the combining effects of a fundamental and third harmonic are shown, with the harmonic in phase with the fundamental. (A harmonic is considered in phase with the fundamental when the instantaneous values of the harmonic and the fundamental pass through zero and go in a positive direction simultaneously.) In Fig. 1-12b the third harmonic is 180° out of phase with the fundamental, but horizontal-axis symmetry still prevails.

(A harmonic is considered to be out of phase with the fundamental by 180° if, at the instantaneous values of zero, the harmonic goes in a negative direction at the time that the fundamental goes in a positive direction.) Shown in Fig. 1-12c is the composite signal formed by combining a fifth harmonic in phase with the fundamental. Since the signals shown in Fig. 1-12 *all* have point symmetry, they are odd-function types and *all cosine terms are zero.*

The addition of an in-phase second harmonic to the fundamental is shown in Fig. 1-13a. Note that the conditions of both point symmetry and horizontal-axis symmetry are met, indicating even-harmonic components and sine terms only. The average value (a_o) is zero.

In Fig. 1-13b, the second harmonic is out of phase by 90° with respect to the fundamental. Now the vertical-axis symmetry indicates only cosine terms are present. In Fig. 1-13c the fifth harmonic leads the fundamental by 90°. Compare this waveform with that in Fig. 1-12c and note that horizontal-axis symmetry is still present, indicating odd-harmonic content.

The waveforms in Figs. 1-11 through 1-13 have a low-order harmonic

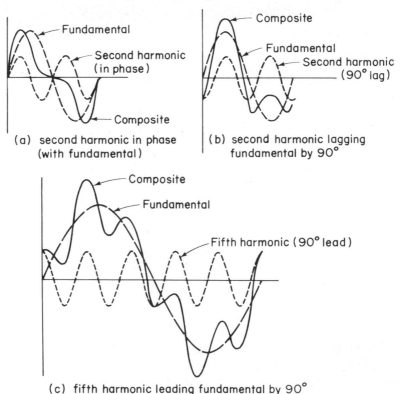

(a) second harmonic in phase (with fundamental)

(b) second harmonic lagging fundamental by 90°

(c) fifth harmonic leading fundamental by 90°

figure 1-13 effect of cosine signals on composite waveform

content. The symmetry principles also apply, however, to more complex waveforms. The square wave shown in Fig. 1-7f, for instance, has horizontal-axis symmetry, indicating odd-harmonic content. As higher-order harmonics are included in the composite waveform, there will be increased steepness of the rising and declining amplitude portions. Hence, in waveform analysis it is important to take into consideration as many harmonic components as possible to get a true representation of the original complex wave. Theoretically, an infinite number of Fourier coefficients can be calculated using the integrals [Eqs. (1-24) and (1-25)] shown earlier. Because of attenuation factors in circuitry, and because of decreasing amplitudes of the higher harmonics (which dilute their significance), a finite number is possible.

1-6. Fourier analysis

A sawtooth waveform, such as shown in Fig. 1-14a, can be represented mathematically by the equation $i = \omega t$, which shows a rising current (i) for $t = 0$ to $t = 2\pi$. Since the signal only has a positive polarity, it is obvious that a dc component (average value) is present. For solving the average value (a_o) Eq. (1-23) is used, substituting ωt for $f(t)$ to represent the signal to be analyzed:

$$a_o = \frac{1}{\pi} \int_0^{2\pi} f(t)\, d(\omega t) = \frac{1}{\pi} \int_0^{2\pi} \omega t\, d(\omega t) = 2\pi$$

Hence

$$\frac{a_o}{2} = \pi$$

Comparing the sawtooth signal of Fig. 1-14a with that of b, it will be noted that moving the horizontal zero line upward to provide for negative and positive polarities produces *point symmetry*. Since the harmonic components in terms of sine and cosine factors are the same for either waveform, the point symmetry indicates that the cosine constants (a_n) are zero and only the sine (b_n) terms are present. This could be proved by substituting ωt for $f(t)$ in Eq. (1-24) and integrating.

Since only the sine terms are involved, Eq. (1-25) is used. Reference can be made to integral tables or the integration performed directly as follows:

$$b_n = \frac{1}{\pi} \int_0^{2\pi} \omega t \sin n\omega t\, d(\omega t)$$

$$= \frac{1}{\pi} \left[\frac{1}{n^2} \sin n\omega t - \frac{1}{n} \omega t \cos n\omega t \right]_0^{2\pi}$$

$$= \frac{1}{\pi} \left[\frac{1}{n^2} \sin 2n\pi - \frac{1}{n}(2\pi) \cos 2n\pi - \frac{1}{n^2} \sin 0 + \frac{1}{n}(0) \cos 0 \right]$$

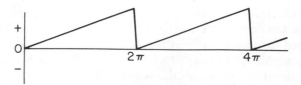

(a) sawtooth waveform containing dc component

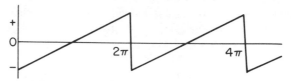

(b) sawtooth waveform with no dc component

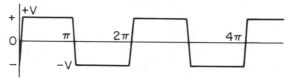

(c) squarewave with no dc component

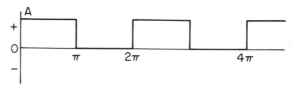

(d) squarewave containing dc component

figure 1-14 sawtooth and squarewave relationships

With n representative of a whole number, we now obtain

$$b_n = -\frac{2}{n}$$

The a_o and b_n values thus obtained are now inserted into the second portion of Eq. (1-21) only, since no cosine terms are present. Because n represents the sequence of harmonics under analysis, $n = 1$ for the first b, 2 for the second, 3 for the third, etc. Thus we find that the sawtooth wave is made up of successive harmonics (both even and odd). Compared to the fundamental, the harmonics have decreasing amplitudes in the order of $\frac{1}{2}, \frac{1}{3}, \frac{1}{4}, \frac{1}{5}$, etc.:

$$i = \pi - 2(\sin \omega t + \tfrac{1}{2} \sin 2\omega t + \tfrac{1}{3} \sin 3\omega t + \tfrac{1}{4} \sin 4\omega t + \cdots)$$

As shown in Fig. 1-8, in the formation of a sawtooth of this type the sinewaves all go in a negative direction at the start of the sawtooth, where $t = 0$. This accounts for the minus sign in the foregoing example. Units could be expressed in v for voltage instead of i for current, as shown. Percentage values are readily ascertained. Thus, if the fundamental is 12 milliamperes (mA), for instance, the second harmonic would be one half, or 6 mA; the third would be 4 mA; the fourth 3 mA, etc.

For the sawtooth waveform shown in Fig. 1-14b, the average value is zero, since the positive and negative values cancel. The harmonic content remains the same as the sawtooth signal of Fig. 1-14a. Thus the series would omit the π for the average value:

$$i = -2(\sin \omega t + \tfrac{1}{2} \sin 2\omega t + \tfrac{1}{3} \sin 3\omega t + \tfrac{1}{4} \sin 4\omega t + \cdots)$$

The general variable x can be used instead of ωt. In such instances the variable x may represent voltage, current, time, etc. Peak amplitude could be represented as A, or voltage as V, as shown in Fig. 1-14c for the square wave. If we use the variable x we can indicate the functions as

$$f(x)_0^\pi = +V \tag{1-26}$$

$$f(x)_0^{2\pi} = -V \tag{1-27}$$

For the square wave in Fig. 1-14c we have zero average value; hence, $a_o = 0$. Note that the square wave is symmetrical about the horizontal axis, indicating only odd harmonics are present, as described earlier. Also, the waveform has point symmetry; hence, only the sine (b_n) terms are present. Integrating, using x and V designations, produces

$$b_n = \frac{1}{\pi} \int_0^\pi V \sin nx \, dx + \frac{1}{\pi} \int_\pi^{2\pi} (-V) \sin nx \, dx$$

$$= \frac{1}{\pi}\left[-\frac{V}{n} \cos nx \right]_0^\pi - \frac{1}{\pi}\left[-\frac{V}{n} \cos nx \right]_\pi^{2\pi}$$

$$= \frac{1}{\pi} \cdot \frac{2V}{n} + \frac{1}{\pi} \cdot \frac{2V}{n}$$

If n is an odd number this produces $b_n = \dfrac{4V}{\pi n}$.

If n is an even number, b_n is zero.

The series then resolves into the following:

$$= \frac{4V}{\pi}\left[\sin x + \frac{1}{3} \sin 3x + \frac{1}{5} \sin 5x + \cdots \right]$$

Thus, we prove that the square wave is composed of a fundamental plus a succession of odd harmonics, with the third harmonic having an amplitude one third of the fundamental, the fifth having one-fifth amplitude, the seventh one-seventh amplitude, etc.

In Fig. 1-14d the horizontal zero reference line has been moved to the bottom of the square-wave train, resulting in a series of periodic rectangular pulses, as shown. Since the time interval from π to 2π is still the same as from 0 to π, the sine and odd-frequency characteristics remain the same as the square wave of Fig. 1-14c. Using A for maximum amplitude instead of the V used in Eqs. (1-26) and (1-27) we have

$$f(x)_0^\pi = A \quad \text{and} \quad f(x)_\pi^{2\pi} = 0$$

From inspection it is obvious that the average dc value from 0 to 2π is

$$a_o = \frac{A}{2}$$

Since only the sine terms are present, the integration is

$$b_n = \frac{1}{\pi} \int_0^\pi A \sin nx \, dx$$

$$= \frac{1}{\pi} \left[-\frac{A}{n} \cos nx \right]_0^\pi$$

$$= \begin{cases} \dfrac{1}{\pi} \cdot \dfrac{2A}{n} & \text{when } n \text{ is an odd number} \\ 0 & \text{when } n \text{ is an even number} \end{cases}$$

$$= \frac{A}{2} + \frac{2A}{\pi} \left[\sin x + \frac{1}{3} \sin 3x + \frac{1}{5} \sin 5x + \cdots \right]$$

Again our Fourier series indicates a fundamental plus a series of odd harmonics with decreasing amplitudes in the order $\frac{1}{3}, \frac{1}{5}, \frac{1}{7}, \frac{1}{9}$, etc.

For the triangular wave shown in Fig. 1-15a, there is a current rise $i = \omega t$ from $t = 0$ to $t = \pi$. From π to 2π the current falls, $i = 2\pi - \omega t$. This wave can be analyzed by using separate integrals for the two parts of the waveform, as was done for the square wave [Eqs. (1-26) and (1-27)]. Thus, the dc average becomes

$$a_o = \frac{1}{\pi} \int_0^\pi \omega t \, d(\omega t) + \frac{1}{\pi} \int_\pi^{2\pi} (-\omega t) \, d(\omega t) = \frac{\pi}{2}$$

The triangular wave with positive and negative values is shown in Fig. 1-15b. Note that from 0 to 2π the waveform has the axis-symmetry characteristics shown in Fig. 1-11a and, hence, contains only cosine values a_n. The Fourier coefficient is split into two parts also, and integration produces an end result indicating only odd-harmonic components.

$$a_n = \frac{1}{\pi} \int_0^\pi \omega t \cos n\omega t \, d(\omega t) + \frac{1}{\pi} \int_\pi^{2\pi} (-\omega t) \cos n\omega t \, d(\omega t)$$

$$= \frac{\pi}{2} - \frac{4}{\pi} \left(\cos \omega t + \frac{1}{9} \cos 3\omega t + \frac{1}{25} \cos 5\omega t + \frac{1}{49} \cos 7\omega t + \cdots \right)$$

The same series is obtained for the waveform shown in Fig. 1-15b

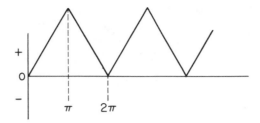

(a) triangular waveform containing dc component

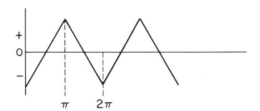

(b) triangular waveform with no dc component

figure 1-15 triangular waveforms

except that the average value is zero. The general variable x could, of course, also be used (with V or A) and the series would become

$$-\frac{4A}{\pi}\left[\cos x + \frac{1}{9}\cos 3x + \frac{1}{25}\cos 5x + \frac{1}{49}\cos 7x + \cdots\right]$$

For the square wave shown in Fig. 1-14c, a constant amplitude $+V$ is present for the first half of the wave and $-V$ for the second half, representative of the *slope of the triangular wave*. Thus, the square wave is actually the differential coefficient of the triangular wave. If we again differentiate the triangular series, term by term, we shall obtain the original series shown for the square wave.

questions and problems

1-1. Explain what is meant by the term *velocity*.

1-2. What factors determine the actual velocity of sound?

1-3. What is the wavelength of a radio wave if its frequency is 120 MHz?

1-4. For Question 1-3, what is the wavelength in feet?

1-5. If the wavelength is 25 m, what is the frequency of the signal?

1-6. If the frequency of a signal is 164 MHz, what is the wavelength in feet?

1-7. For Question 1-6, what is the wavelength in centimeters?

1-8. How many angstroms are there in 1 meter?

1-9. Give the frequency range in megaherz of the following abbreviated designations: MF, VHF, and SHF.

1-10. Define the terms *carrier* and *sidebands*.

1-11. Define the terms *modulation* and *harmonics*.

1-12. What sound characteristics provide the distinguishing features that permit recognition of individual musical instrumental tones?

1-13. How are harmonic signals related to the fundamental signal frequency and amplitude?

1-14. What is the purpose for using the Fourier series?

1-15. What are the coefficients of the Fourier series?

1-16. If peak amplitude (A) is 200, what is the instantaneous amplitude (a) at the 45° angle?

1-17. For the waveform shown in Fig. 1-16a, define the symmetry and ascertain whether it contains odd, even, or both odd and even harmonic components.

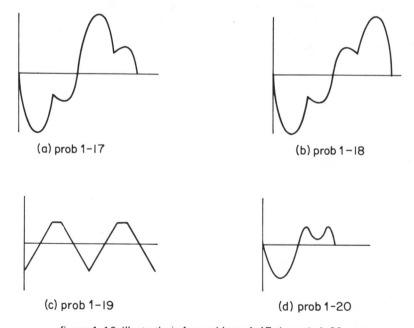

(a) prob 1-17

(b) prob 1-18

(c) prob 1-19

(d) prob 1-20

figure 1-16 illustrations for problems 1-17 through 1-20

1-18. What harmonic components are present in the signal shown in Fig. 1-16b and are sine or cosine terms present?

1-19. Define the symmetry for the signal shown in Fig. 1-16c and indicate which terms (sine or cosine) are present.

1-20. For the signal shown in Fig. 1-16d, what harmonic component is present and what is its phase relationship to the fundamental?

1-21. Prove, by integration, that the Fourier coefficient given below for the cosine function of a sawtooth waveform yields zero:

$$a_n = \frac{1}{\pi} \int_0^{2\pi} \omega t \cos n\omega t \, d(\omega t)$$

1-22. A waveshape has a rising amplitude from 0 to π, where $i = \omega t$. From π to 2π, however, i has a constant amplitude: $i = \pi$. Hence, the Fourier coefficient is

$$a_n + b_n = \frac{1}{\pi} \int_0^{\pi} \omega t \cos n\omega t \, d(\omega t) + \frac{1}{\pi} \int_\pi^{2\pi} \pi \sin n\omega t \, d(\omega t)$$

Solve for average values and complete the series.

1-23. Explain why the square-wave signal is actually the differential coefficient of the triangular wave.

2

modulation and sidebands

2-1. introduction

The functional aspects of modulation circuits and systems can be learned more readily by first becoming thoroughly familiar with the basic characteristics of signal waveforms. In Chapter 1 the harmonic principles related to complex waveforms were introduced, and it was shown that signals having frequencies other than the fundamental were present when the signal contained deviations from that of a pure sinewave. This factor also relates to the sideband signals produced during both AM and FM processes.

Thus, this chapter supplements the material of Chapter 1 and presents the final data for a complete understanding of the in-circuit characteristics of modulated signals. During propagation, all the signals comprising the composite modulation waveform relate to the information to be conveyed. Hence, if all are transmitted, full efficiency is procured and high fidelity assured. Some signals may be eliminated prior to the transmission (for instance, single sideband systems) to conserve spectrum space, though overall efficiency declines, as more fully described later.

Though the transmitted signals differ electrically from the in-circuit types, as discussed in Chapter 12, the harmonic signal content, sideband relationships, modulation amplitudes, and other factors coincide with the in-circuit signals present in the final stages of the transmitter. Thus, the material in this chapter serves as a foundation for the study of modulation circuits and transmission principles involving the various branches of modern communication systems. All aspects of the practical circuitry (which begins with Chapter 3) are influenced by the signal characteristics discussed in Chapters 1 and 2.

2-2. AM signals

As described in Section 1-4, to transmit audio, video, or other low-frequency signals, they must be used to modulate a high-frequency signal capable of spanning distances far in excess of those spanned by low-frequency waveforms. There are various methods for modulating a high-frequency (carrier) signal with a low-frequency signal. One of the earliest processes and one still widely used is that known as *amplitude modulation* (AM). This system is employed in radio broadcasting (550 kHz–1700 kHz) and is the modulation process used for the picture in television. [The sound portion of television is *frequency modulated* (FM)].

Basically, AM is a modulation method wherein the high-frequency carrier signal is modified so that the amplitude varies to correspond to the amplitude variations of the low-frequency modulating signal. Actually, the audio or other low-frequency signal is not "carried" by the transmitted signal, and is represented only by corresponding changes in the amplitude of the modulated RF signal. As described in Chapter 6, a detector circuit sensitive to amplitude changes produces audio or other low-frequency signals closely resembling the original modulating signals.

A basic AM radio transmitter is shown in Fig. 2-1. The sound input is amplified to the necessary degree by preamplifiers and power stages until the required level is reached. An oscillator generates the RF carrier signal, which is amplified by several Class C circuits, until the needed output power of the

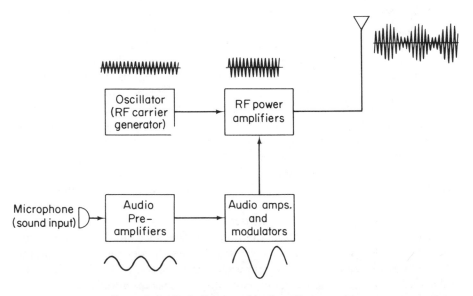

figure 2-1 block diagram of basic radio transmitter

carrier has been reached. In most commercial transmitting stations the carrier power is stepped up until many *kilowatts* of power are radiated. High carrier power increases the signals arriving at the detection systems of remote receivers and also increases the signal-to-noise ratio of reception.

As shown in Fig. 2-1, the amplified audio signal is combined with the

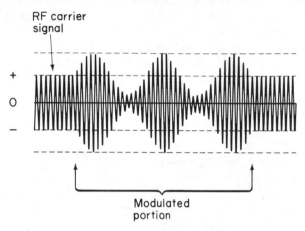

figure 2-2 amplitude modulation (AM)

amplified RF carrier signal, and the resultant is a carrier that has amplitude variations which correspond to the audio-modulating signal. It is this waveform that is sent to the antenna system for broadcasting.

As shown in Fig. 2-2, the modulated portion of the carrier signal varies in amplitude above and below the unmodulated carrier amplitude. Also note that the carrier is essentially an ac signal with positive and negative alternations, but because of its high frequency it is termed an RF signal.

In Fig. 2-3 the audio-modulating signals are compared to the resultant modulated carrier. In Fig. 2-3a, the audio signal has a fairly high amplitude and the result (shown in Fig. 2-3b) is that the carrier amplitude increases and decreases to a greater degree than would be the case for a lower-amplitude audio-modulating waveform. With a higher-frequency audio signal the carrier amplitudes would change more rapidly to conform to the audio signals. With a lower-amplitude modulated carrier, as shown in Fig. 2-3d, the same principles apply, except a lower-amplitude audio-modulating signal (Fig. 2-3c) is needed to achieve the same degree of modulation that would be required for a higher amplitude carrier.

As the amplitude of the modulating signal is increased, the degree of modulation also increases. For practical purposes the degree of modulation is usually expressed in percentages, and for 100 per cent modulation the waveform appears as shown in Fig. 2-4a. Note that the peak amplitude is twice that of the unmodulated carrier, and the minimum amplitude is virtually zero.

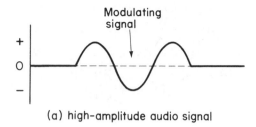

(a) high-amplitude audio signal

(c) low-amplitude audio signal

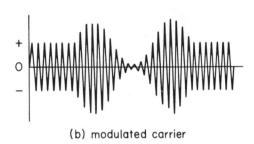

(b) modulated carrier

figure 2-3 degrees of amplitude
modulation

(d) modulated carrier

The *degree of modulation m* is expressed as

$$m = \frac{E_{\max} - E_{\min}}{E_{\max} + E_{\min}} \qquad (2\text{-}1)$$

Multiplying the results of Eq. (2-1) by 100 gives the *percentage* of modulation. If the modulation exceeds 100 per cent, portions of the modulating information are lost and signal distortion is present during the demodulation process at the receiver. For overmodulation, the carrier amplitude rises to

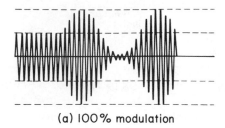

(a) 100% modulation

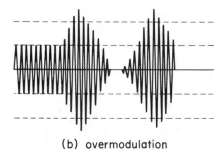

(b) overmodulation

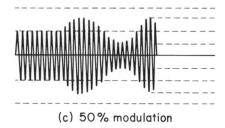

(c) 50% modulation

figure 2-4 amplitude modulation percentages

more than twice that of the carrier and drops to zero level, remaining there for a perceptible interval, as shown in Fig. 2-4b.

The degree of modulation does not depend on carrier amplitude alone, as is evident from Eq. (2-1), but on the relative proportions of minimum and maximum waveform values of either carrier voltage or carrier current. If we have a peak value of 750 volts (V) and a minimum value of 250 V, the percentage of modulation is 50:

$$\frac{750 - 250}{750 + 250} = 0.5 \times 100 = 50 \text{ per cent}$$

As shown in Fig. 2-4c (for a sinewave change of amplitude), the carrier would be 500 V. For a 1000 V carrier with a maximum peak of 1500 V and a minimum of 500 V, the percentage of modulation remains the same (50 per cent).

In Section 1-4 it was pointed out that a pure ac signal has only a single frequency, but that any distortion of the waveform creates signals of frequencies other than the fundamental frequency. Thus, when we cause the carrier signal to change in amplitude, it undergoes a form of waveform distortion. This is evident by inspection of a random selection of any single cycle from among those that are undergoing an amplitude change, as shown in Fig. 2-5. Note that one alternation differs in amplitude from the other for

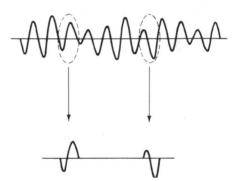

figure 2-5 sinewave distortion during modulation

successive cycles, indicating waveform distortion of the carrier. The continuous carrier distortion during the modulation process generates other signals; hence, the modulated carrier can no longer be considered a single frequency, but instead a *composite* signal made up of several signals with frequencies differing from each other. The related mathematics indicates the nature of the signals formed.

From Eq. (1-15) we found that an unmodulated sinewave signal is expressed as

$$e = E \sin \omega t$$

If we cause the amplitude E to vary sinusoidally at a frequency rate below the carrier, the instantaneous voltage becomes

$$E = E_c(1 + m \sin \omega_c t) \qquad (2\text{-}2)$$

where $E =$ maximum voltage at any instant

E_c = peak amplitude of the unmodulated carrier signal
ω_c = angular velocity of the carrier signal
m = degree of modulation [Eq. (2-1)]

Combining this with Eq. (2-2), we obtain

$$e = E_c(1 + m \sin \omega_m t) \sin \omega_c t \qquad (2\text{-}3)$$

where e = instantaneous amplitude of wave
E_c = peak amplitude of carrier signal
m = degree of modulation
ω_m = angular velocity of modulating signal
ω_c = angular velocity of carrier signal

When the right-hand side of Eq. (2-3) is multiplied out, we obtain

$$e = E_c \sin \omega_c t + m E_c \sin \omega_m t \sin \omega_c t \qquad (2\text{-}4)$$

The last term may now be expanded into functions of sum and difference angles by the trigonometric formula $\sin x \sin y = \frac{1}{2}[\cos (x - y) - \cos (x + y)]$. Thus, the equation of a carrier with sinewave-signal modulation becomes

$$e = E_c \sin \omega_c t + \frac{mE_c}{2} \cos(\omega_c - \omega_m)t - \frac{mE_c}{2} \cos(\omega_c + \omega_m)t \qquad (2\text{-}5)$$

Inspection of Eq. (2-5) indicates that the modulated wave consists of three individual signals, one of which is the original carrier wave ($E_c \sin \omega_c t$). The other components are called *sideband* signals. For a given sinewave modulating signal, two sidebands are produced. One (the upper sideband) has a frequency above the carrier signal equal to the carrier frequency plus the modulation signal frequency [$(mE_c/2) \cos(\omega_c + \omega_m)t$]. The other (lower sideband) has a frequency lower than the carrier by a frequency equal to the modulating signal frequency, as shown in Fig. 2-6. For 100 per cent modulation by a single ac modulating signal, each of the sidebands has an amplitude one half that of the unmodulated carrier.

The sideband signals contain the modulated information and the sideband amplitudes relate to the amount of modulation (the *volume* for audio signals). The *carrier alone (without sidebands) contains no amplitude changes* and hence conveys no modulating information. The modulating information can be transmitted by using the carrier and only a single sideband. This procedure narrows the spectrum span (called *bandwidth*) occupied by the composite waveform. The carrier signal could also be omitted and only a single sideband transmitted. In such an instance the missing carrier signal (of proper frequency) must be furnished at the receiving end for proper demodulation procedures.

In the modulation by a complex waveform (speech or video signals) many sidebands are formed, each distant from the carrier by the frequency of

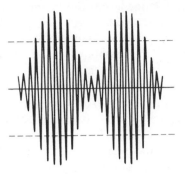

(a) amplitude modulation
(carrier plus sidebands)

(b) carrier only
(less sidebands)

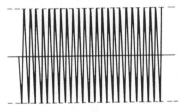

(c) upper sideband only

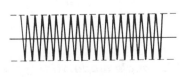

figure 2-6 signals in modulated carrier   (d) lower sideband only

the modulating signal producing it. Undesired signals (spurious harmonics) are also usually present in the complete transmitting system, and such unwanted signals must be filtered out as much as possible before the modulated energy is finally applied to the transmitting antenna. These factors are discussed more fully later.

For Fig. 2-6 the carrier and sideband signals cannot be represented exactly since the high frequencies involved in these signals prohibit indicating the actual number of cycles that occur in a given time interval. If, for instance, the carrier is 800 kHz and the audio-modulating signal is 500 Hz, the carrier amplitude increases and decreases 500 times per second. Because there are 800,000 cycles of carrier signal per second, there would be 1600 carrier cycles for every audio cycle. Thus, one cycle of the 500-Hz audio signal would cause a modulating increase and decrease of the carrier amplitude that would span 1600 cycles of the carrier.

With an 800-kHz carrier modulated by a 500-Hz signal, the sideband signals will have frequencies of 800.5 kHz and 799.5 kHz. Thus, it is evident that a carrier, plus its upper and lower sidebands, requires more spectrum space than an unmodulated carrier. For the 500-Hz modulating signal there is a frequency span of 1 kHz between the lower and upper sideband, requring such a spectrum span to accommodate the band of signals involved. The carrier frequency does not determine the bandwidth in AM, since a 2000-kHz carrier, modulated by the same 500-Hz audio signal, would have the same bandwidth as the 800-kHz carrier. If the modulating frequency is 5 kHz, the two sidebands, plus the carrier, would require a 10-kHz-wide span, or bandwidth, as shown in Fig. 2-7.

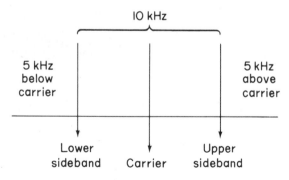

figure 2-7 AM radio bandpass

2-3. angle modulation

Consider the wave equation

$$e = A \cos(\omega t + \theta) \qquad (2\text{-}6)$$

Here A represents the maximum amplitude of either the voltage or current of the signal, and $\omega t + \theta$ is the instantaneous angle and phase of the function.

For such a wave the modulation process can consist of varying three things: the amplitude (A); the frequency (involving the angular velocity ωt), or the phase angle θ. When the modulating signal alters either the frequency or phase of the carrier signal, the system is referred to as *angle* modulation. Both the frequency- and phase-modulation systems of angle modulation are essentially similar, as are the demodulating processes. Hence, an FM receiver will process both FM and PM equally.

A typical frequency-modulated wave is shown in Fig. 2-8. Here the

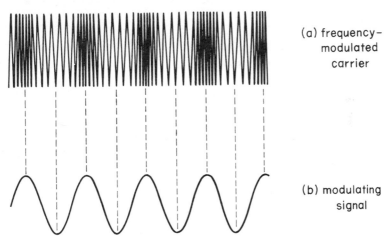

(a) frequency-modulated carrier

(b) modulating signal

figure 2-8   frequency modulation (FM)

carrier frequency is shifted above normal frequency (called *resting* frequency), as well as below the carrier frequency, at a rate dependent on the frequency of the modulating signal. The amount of shift depends on the amplitude of the modulating signal. Hence, a higher volume of audio-modulating signal, for instance, will cause a greater shift in carrier from its resting frequency than would a lower-volume audio signal.

When, however, we vary the *frequency* of the carrier *directly* during the modulation process (known as *direct* FM), the phase is also varied, though indirectly. Similarly, if we vary the phase directly during the modulation process, the frequency is also varied, though indirectly.

For frequency modulation (FM), the ratio of the frequency shift (deviation) of the carrier to the frequency of the modulating signal causing that carrier deviation is called the *modulation index* (not to be confused with the *deviation ratio* described later). In equation form the modulation index becomes

$$m_f = \frac{\Delta f_c}{f_m} \quad \text{or} \quad m_f = \frac{k_f E_m}{\omega_m} \tag{2-7}$$

where m_f = modulation index for FM

Δf_c = frequency deviation of carrier

f_m = frequency of modulating signal

k_f = degree of frequency variation as a function of the amplitude of the modulating signal

E_m = amplitude of modulating signal

ω_m = angular velocity of modulating signal

For phase modulation (PM), the modulation index is expressed in equation form as

$$m_p = k_p E_m \tag{2-8}$$

where m_p = modulation index for PM

k_p = degree of phase variation as a function of the amplitude of the modulating signal

E_m = amplitude of modulating signal

Thus, for angle modulation, the classification is frequency modulation when the modulation index is inversely proportional to the frequency of the modulating signal. For phase modulation, the modulation index is independent of the frequency of the modulating signal.

The *deviation ratio* is based on maximum values as opposed to the instantaneous characteristic of the modulation index. The deviation ratio relates the maximum deviation of the carrier frequency to the highest frequency modulating signal:

$$\text{deviation ratio} = \frac{\text{maximum frequency deviation of carrier}}{\text{highest frequency of modulating signal}} \tag{2-9}$$

When a deviation ratio value is assigned to an FM station, it establishes the limits of the spectrum span occupied by the carrier shift relative to the audio frequency of the signals causing such carrier shift. It must be remembered, however, that a station does not operate under continuous maximum conditions. Audio amplitudes vary considerably; hence, the amount of carrier shift changes during the entire broadcast. Similarly, the frequency of the audio signals changes constantly, producing a corresponding change in the rate of the carrier shift above and below the resting frequency of the FM carrier.

2-4. FM and PM factors

Equation (2-6) can be expressed as $A \cos \phi$, where $\phi = \omega_c t + \theta$ and ω_c is the angular velocity of the unmodulated carrier signal. The angular velocity can be shown as

$$\omega = \frac{d\phi}{dt} \qquad (2\text{-}10)$$

The *modulating signal* in FM is expressed as

$$e_m = E_m \cos \omega_m t \qquad (2\text{-}11)$$

Equations (2-10) and (2-11) provide us with the following equation showing a waveform that is varied in frequency by the angular velocity according to the amplitude of the modulating signal:

$$\omega = \omega_c + k_f E_m \cos \omega_m t \qquad (2\text{-}12)$$

The k_f and E_m expressions, as well as others, were given earlier for Eq. (2-7). The instantaneous voltage is shown as

$$e = E_c \sin \left[\int (\omega_c + k_f E_m \cos \omega_m t)\, dt + \theta_0 \right] \qquad (2\text{-}13)$$

where θ_0 is the *initial* phase angle and can be considered as $\theta_0 = 0$, since it has no influence on the frequency modulation. Integrating, we obtain

$$e = E_c \sin \left(\omega_c t + \frac{k_f E_m}{\omega_m} \sin \omega_m t \right) \qquad (2\text{-}14)$$

Since one portion is the frequency-modulation index m_f [Eq. (2-7)], the formula for the FM wave becomes

$$e = E_c \sin(\omega_c t + m_f \sin \omega_m t) \qquad (2\text{-}15)$$

If the modulating signal [Eq. (2-11)] is expressed as

$$e_m = E_m \sin \omega_m t \qquad (2\text{-}16)$$

the formula for the FM wave then is

$$e = E_c \sin(\omega_c t - m_f \cos \omega_m t) \qquad (2\text{-}17)$$

When phase modulation is employed, the instantaneous change in phase (θ) is produced by the modulating signal; hence

$$\theta = \theta_0 + m_p \sin \omega_m t \qquad (2\text{-}18)$$

where m_p is the modulation index for phase modulation and is $k_p E_m$, as shown earlier for Eq. (2-7). Again, with $\theta_o = 0$, the formula for the phase-modulated (PM) wave is

$$e = E_c \sin(\omega_c t + m_p \sin \omega_m t) \qquad (2\text{-}19)$$

Compare Eq. (2-19) with (2-15) and note that only the modulating index differs. As explained earlier, in one instance (FM) the modulation index is inversely proportional to the frequency of the modulating signal, while for the other (PM) the index is independent of the modulating-signal frequency.

In contrast to AM, where a single sinewave modulating signal only produces two sidebands, many sidebands are formed in frequency modulation for a single modulating wave. Theoretically, an infinite number of sidebands is generated, but only a few have sufficient amplitude to be significant in terms of bandwidth or the detection process. Such sideband factors become evident through mathematical procedures involving the expansion of Eq. (2-15) by what are known as *Bessel functions*. The term $J_n(x)$ designates the Bessel function (of x) of the first kind of order n. The factor x becomes m_f for frequency modulation and m_p for phase modulation, as discussed for the modulation index. The results of such expansion yield the following:

$$
\begin{aligned}
e_o = E_c\{ & J_0(m_f) \sin \omega_o t + J_1(m_f)[\sin (\omega_o + \omega_m)t - \sin (\omega_o - \omega_m)t] \\
& + J_2(m_f)[\sin (\omega_o + 2\omega_m)t + \sin (\omega_o - 2\omega_m)t] \\
& + J_3(m_f)[\sin (\omega_o + 3\omega_m)t - \sin (\omega_o - 3\omega_m)t] \\
& + J_4(m_f)[\sin (\omega_o + 4\omega_m)t + \sin (\omega_o - 4\omega_m)t] \\
& + J_5(m_f) \cdots \}
\end{aligned}
\tag{2-20}
$$

The first term represents the carrier; the others designate sidebands. Carrier and sideband amplitudes vary with the modulation index (m_f or m_p). Several of the functions are graphed in Fig. 2-9 and show carrier versus

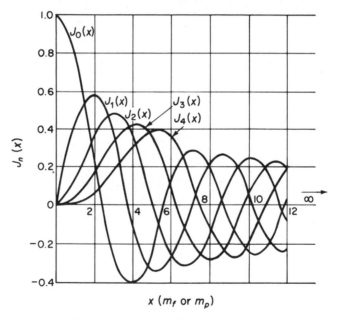

figure 2-9 curves of Bessel functions

sideband magnitudes for any modulation index up to 12. Note the changes in relative amplitudes of sidebands and carrier for various values of the modulation index. When the modulation index is 2, for instance, the first two sets of sidebands have a much greater amplitude than the carrier. At a modulation index of approximately 2.4, the carrier amplitude reaches 0, and all the transmitted power is contained in the sidebands only. Figure 2-9 indicates that the carrier changes amplitude for changes in the modulation index. The data obtained from the Bessel functions can be tabulated as shown in Table 2-1 shown on page 42.

Blank sections in the table are sideband amplitudes omitted because they no longer have significant values. For an unmodulated wave, $J_0(x)$ maximum carrier amplitude is designated as 1.0, though the algebraic addition of carriers and sidebands for the various values of the modulation index will not produce 1.0. Since ac waveforms are involved, vector addition is necessary (the square root of the sum of squares for the values given).

In using Table 2-1 we evaluate the horizontal representations of the sidebands with respect to the carrier for a given modulation index. If, for instance, the m_f is 1, the carrier $[J_0(x)]$ will have an amplitude of $0.765E_c$, where $E_c = 1$ represents the maximum magnitude of the unmodulated carrier. The first-order sidebands have a magnitude of $0.44E_c$ each, and the frequency of the upper $J_1(x)$ sideband is $f_c + f_m$, where f_c is the carrier frequency and f_m the *modulating-signal* frequency. The lower sideband $J_1(x)$ is $f_c - f_m$.

The magnitude of each of the second-order sidebands is $0.115E_c$, with the upper sideband $f_c + 2f_m$ and the lower $f_c - 2f_m$. The third-order sidebands are the last of significant value when $m_f = 1$, and the amplitude is $0.002E_c$. Frequencies are $f_c + 3f_m$ and $f_c - 3f_m$.

Even though additional sets of sidebands are produced for higher values of the modulation index, all the sidebands in FM are spaced from *each other* by a frequency equal to the modulating-signal frequency. Thus, if the modulating signal has a frequency of 2 kHz, each sideband is spaced from the other by 2 kHz. The *rate* of the carrier shift is also determined by the modulating-signal frequency, and the higher the modulating frequency the more rapid the frequency excursions of carrier each side of central (resting) frequency. The *amplitude* of the modulating signal controls the *extent* of the frequency shift of the carrier.

For a modulation index of 0.4 or less, only one significant sideband is produced above and below the carrier. For a modulation index of 0.5, two sidebands are generated above the carrier and two below. Table 2-2 indicates the number of significant sidebands that exist for a modulation index range from 1 to 10 (see page 43).

In standard FM broadcasting for the general public (between 88 MHz and 108 MHz) the maximum deviation is 75 kHz, and the highest audio frequency used for modulating purposes is 15 kHz. Thus, the deviation ratio [Eq. (2-9)] is

$$m_f = \frac{75 \text{ kHz}}{15 \text{ kHz}} = 5$$

In both VHF and UHF television transmission the FM sound portion

table 2-1

modulation index (m)	carrier amplitude $J_0(x)$	relative amplitude of sidebands								
		$J_1(x)$	$J_2(x)$	$J_3(x)$	$J_4(x)$	$J_5(x)$	$J_6(x)$	$J_7(x)$	$J_8(x)$	$J_9(x)$
0	1.000									
0.01	1.000	0.005								
0.02	0.999	0.010								
0.05	0.999	0.025								
0.1	0.998	0.050								
0.2	0.990	0.100								
0.5	0.938	0.242	0.310							
1.0	0.765	0.440	0.115	0.003						
2.0	0.224	0.577	0.353	0.129	0.034					
3.0	−0.260	0.339	0.486	0.309	0.132	0.043	0.012			
4.0	−0.397	−0.066	0.364	0.430	0.281	0.132	0.049	0.015		
5.0	−0.178	−0.328	0.047	0.365	0.391	0.261	0.131	0.053	0.018	
6.0	0.151	−0.277	−0.243	0.115	0.358	0.362	0.246	0.130	0.057	0.021

table 2-2

modulation index	significant sidebands around carrier	
	above	below
0.4	1	1
0.5	2	2
1.0	3	3
2.0	4	4
3.0	6	6
4.0	7	7
5.0	8	8
6.0	9	9
7.0	11	11
8.0	12	12
9.0	13	13
10.0	14	14

uses a maximum deviation each side of the carrier of 25 kHz, resulting in fewer significant sidebands than produced for standard FM broadcasting.

The deviation ratio is actually only indicative of maximum modulating conditions, and since such peak magnitudes occur only rarely in standard broadcasting, the ratio does not represent normal transmission factors. In the broadcasting of music most of the *fundamental* audio frequencies lie below 5 kHz. Signals with higher frequencies are the *overtones* produced when musical instruments are played. These overtones (also called *harmonics*) have a much lower *amplitude* than the fundamental signal frequency and, hence, do not cause as great a deviation of the carrier frequency. Also, since the overtones have higher frequencies, they would be spaced far from the carrier and have low amplitudes.

With a modulation index of 5 there are eight significant sidebands each side of the carrier. Thus, a high musical tone of 4 kHz would produce a total of 16 sidebands, each spaced from the other by 4 kHz, with a total bandwidth of 4 kHz $\times$ 16 = 64 kHz. For lower-frequency musical tones the spread of the significant sidebands would, of course, be much less. The harmonics of musical tones, having much lower amplitudes than the fundamentals, would not cause as great a carrier deviation and, hence, have a modulation index value below the maximum allocated. Consequently, there would be fewer significant sidebands and, hence, little risk of causing the bandwidth to exceed the maximum permitted deviation.

Thus, during normal FM transmission, the 75-kHz limit set for deviation on each side of the central carrier frequency is adequate in terms of bandwidth for the standard FM broadcast band. As an added safety measure, however, an allocation of an additional 50 kHz was added to each FM station channel (25 kHz on each side) to guard against undesired signals spilling over into adjacent station allocations and causing interference in reception.

The total allocated bandwidth for a standard FM broadcast is shown in Fig. 2-10. The 25 kHz allocated on each side of the maximum deviation are

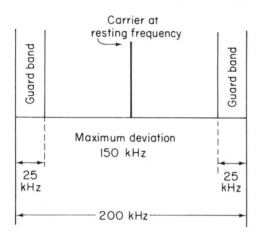

figure 2-10 FM radio frequency allocation

termed *guard bands*, as shown, and they bring the total allocation to 200 kHz.

Note that a modulation index of less than 0.5 in Table 2-1 produces only two significant sidebands for a given sinewave modulating signal (one above and one below the carrier), and hence compares with the two sidebands produced by amplitude modulation. A low deviation ratio is used by many specialized services, such as police, amateurs, and governmental agencies. When FM transmission is such that the bandwidth produced is no wider than for AM for the same modulating-signal frequency, the transmission is known as *narrow-band* FM (NFM). Generally, narrow-band FM has the advantages of occupying less room in the broadcast spectrum, though the system does not have the high efficiency inherent to the systems using a higher modulation index.

In FM the term *degree of modulation* is also used, though "100 per cent modulation" does not have the same meaning as in AM, in which the carrier amplitude varies between zero and two times its normal unmodulated amplitude. In FM it is virtually impossible to overmodulate, so the degree of modulation has been set in relation to the maximum permitted deviation. If the maximum deviation is set at 75 kHz each side of the resting frequency, it is referred to as 100 per cent modulation.

questions and problems

2-1. What type of modulation is used for the picture portion of TV? For the sound portion?

2-2. During the modulation process in an AM transmitter, the carrier current reached 700 amperes (A) maximum and 300 A minimum. What is the percentage of modulation?

2-3. For Problem 2-2, what is the percentage of modulation if the carrier voltage reaches a maximum of 9000 V and a minimum of 1000 V?

2-4. When a signal has a changing amplitude, why are signals of other frequencies produced?

2-5. Rewrite Eq. (2-5) so that its derivation is from $e = E \cos \omega t$ instead of $e = E \sin \omega t$.

2-6. An AM carrier is modulated by three audio signals, one having a frequency of 50 Hz, another of 500 Hz, and a third of 5 kHz. How many sidebands are produced?

2-7. For Problem 2-6, what is the total bandwidth?

2-8. For Problem 2-6, what are the sideband frequencies if the carrier is 800-kHz?

2-9. Why does an FM receiver process phase modulation as well as it does frequency modulation?

2-10. What characteristic of the modulating signal causes a greater frequency shift of an FM signal and what causes a more rapid change each side of the resting frequency?

2-11. Theoretically, the number of sidebands generated during FM reaches infinity. Why, then, is the bandwidth in FM not prohibitive? Explain briefly.

2-12. In FM, what determines the frequency spacing of the sidebands with respect to each other and the carrier?

2-13. Why does the carrier amplitude change during the frequency-modulation process?

2-14. The maximum deviation permitted for a certain FM station is 60 kHz, and the highest modulating signal frequency is 10 kHz. What is the deviation ratio and how many significant sidebands are produced?

2-15. For Problem 2-14, which sidebands have a relative amplitude over twice that of the carrier?

2-16. What are *guard bands* and what is their purpose?

2-17. Describe *narrow-band* FM (NFM) and explain its advantages, if any, over conventional FM.

2-18. Explain how the *degree of modulation* relates to FM and the term 100 per cent modulation.

2-19. If an FM station is permitted a maximum deviation of 50 kHz each side of the carrier resting frequency and the bandwidth allocation is 175 kHz, what is considered 100 per cent modulation for this station?

2-20. An FM station is permitted to deviate a maximum of 30 kHz each side of the carrier, with a maximum frequency of 15 kHz for the audio-modulating signal. Hence, the deviation ratio is 2. However, a deviation of 20 kHz and an audio signal of 10 kHz also give a ratio of 2. Is the second condition also considered 100 per cent modulation? Explain your answer.

3

signal generation and shaping

3-1. introduction

The carrier signal in communication systems is generated by an RF oscillator (see Fig. 2-1). The fundamental frequency of the carrier signal may be established by the oscillator, or the oscillator signal output may have a frequency below that of the transmitted carrier, with frequency multipliers raising the oscillator signal frequency.

High-frequency oscillators are also used in the tuning stages of AM, FM, and TV receivers. On occasion a receiver may use an oscillator to generate the required carrier signal frequency in single sideband-type demodulation systems.

High-frequency oscillators designed to generate sinewave signals are of the *resonant* type, so named because series or parallel coil-capacitor combinations are turned to resonance with the desired frequency. Crystal control as well as special circuits are usually employed to obtain the required stability so that the carrier does not drift from its assigned spectrum space.

Nonresonant oscillators are also widely used, with primary applications including the generation of low-frequency signals from 30 Hz to 20 kHz in the audio spectrum. On occasion, signals with frequencies as high as 500 kHz are generated by the nonresonant oscillators. Such nonresonant oscillators are referred to as *relaxation* types, and the frequency of the generated signals is determined by the selection of resistance and capacitance values of the circuitry. Instead of sinewave signals, the relaxation oscillators produce special waveshapes that are easily converted to the desired shape (sawtooth, pulse, etc.) by special circuits added to the output of the oscillators. Such added circuits modify the output signal as required, and include discharge circuits, clippers, differentiators, and others, as discussed later.

The relaxation-type oscillators permit simple synchronization of the generated frequency by an external signal's frequency, thus providing for exact frequency control in such applications as vertical and horizontal sweep signal synchronization in television, pulse-rate timing in radar, and other applications (discussed in subsequent chapters). Either transistors or vacuum tubes can be used in both the resonant and relaxation oscillator types.

3-2. crystal control

Thin slabs cut from quartz crystals are widely used to stabilize the frequency of RF oscillators to prevent the drift inherent to resonant-circuit signal generators. Quartz crystals have a *piezoelectric* effect and under pressure or strain generate a voltage, thus acting as transducers for converting mechanical energy to electrical. If a voltage is applied across the crystal slab, a distortion occurs in its shape, thus again forming a transducer that now converts electric energy to mechanical.

A typical application of a crystal oscillator is shown in Fig. 3-1a. Here the 3.58-MHz system generates the color subcarrier in a television receiver. In color transmission this subcarrier is suppressed, and only sideband information is transmitted. Hence, the missing carrier must be replaced during the demodulation process (see Chapter 9).

As shown in Fig. 2-1, oscillators are also used to generate the carrier frequency. For a high order of stability, crystal control is utilized, as shown in Fig. 3-1b. A crystal having a frequency lower than the final carrier may be used, with subsequent frequency-multiplying systems increasing the frequency to that desired. A *crystal oven* may be used wherein a heating element (*R*) is regulated to maintain a constant temperature in the heat-insulated chamber. Crystals that have a special cut are less affected by temperature; hence, the temperature control can be dispensed with in services when critical frequency control is not essential.

When specific carrier frequencies are allocated by the Federal Communications Commission (FCC), such as in AM, FM, and TV entertainment-type broadcasting, the stations must keep carrier deviation within limits as prescribed by the FCC. When many individuals or services share a band of frequencies, such as the citizens' band, amateur radio, etc., carrier drift is not as serious nor is operation at a specific frequency mandatory. Precautions must, however, be taken when operation is near the low- and high-frequency limits of the band. Crystal control may also be used to maintain station frequency at an identifiable place in the spectrum for greater reliability in establishing communication.

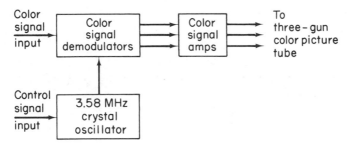

(a) 3.58 MHz crystal oscillator for color subcarrier

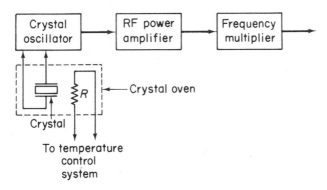

(b) crystal oscillator for carrier frequency

figure 3-1 crystal oscillators in color receivers

As shown in Fig. 3-2, a quartz crystal has a hexagonal cross section with a pointed end. The axis running lengthwise through the center of the crystal and through the end point is known as the Z axis, or *optical* axis. Pressures applied along the Z axis produce no piezoelectric effects. For the cross section, an axis passing through the corners of the hexagon, as shown in Fig. 3-2, is called the X (or electrical) axis. Obviously, in a hexagon, three such axes exist. Three Y axes also are present, as also shown. These are sometimes referred to as the *mechanical* axes.

Crystal slabs are cut at specific angles to the X, Y, or Z axis to obtain required characteristics for resonant frequency, temperature coefficient, and the degree of harmonic signal generation. Thus, a right-angle cut with respect to the X axis forms an X-cut crystal, which has a negative temperature coefficient. Thus, an increase in temperature produces a decrease in the oscillating frequency of the signal produced. The Y cut has a positive temperature coefficient, producing a rise in frequency with an increase in temperature.

When slabs are cut with the face at a specific angle to the Z axis, special types are formed, with designations *AT, BT, CT, DT, ET, FT,* and *GT*. Such crystals are characterized by zero (or almost zero) temperature coeffi-

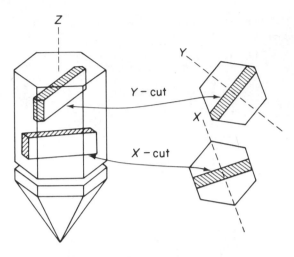

figure 3-2 piezo-quartz crystal cuts

cients. The *AT* cut is extensively used and has a zero temperature coefficient from 40 to 50°C, with operational frequency characteristics from approximately 500 kHz to 10 MHz. A thinner cut produces a correspondingly higher frequency.

The *GT* cut has an exceptionally wide zero temperature coefficient range, 10 to 100°C, and the best frequency operation is around the 100-kHz region. The *ET* and *FT* cuts are primarily harmonic types and produce output signal voltages from the third through seventh harmonic of the fundamental cut. Since the *X* and *Y* cuts have a high temperature coefficient, they are now obsolete types.

The resonant frequency of the crystal is related to the thickness by

$$f = \frac{k}{t} \tag{3-1}$$

where k = constant dependent on type of cut
  f = frequency in kilohertz
  t = thickness in inches

If frequency in Eq. (3-1) is taken in megahertz, the thickness is in thousandths of an inch. The variation in thickness for a specific frequency is evident for the various cuts when considering the difference in k. For the *X* cut, k reaches a high of 112.6, while the *AT* cut has a k of only 66.2. The *Y* cut has a k of 77.0 and the *BT*, 100.78.

Equation (3-1) is useful for ascertaining the frequency of a crystal by measuring the thickness with micrometer calipers. Thus, if the thickness is 0.05039 inch (in.) and the cut is *BT*, the frequency is 2000 kHz, or 2 MHz:

$$f = \frac{k}{t} = \frac{100.78}{0.05039} = 2000 \text{ kHz}$$

During the grinding of a crystal for a specific frequency, the following variation of Eq. (3-1) applies:

$$t = \frac{k}{f} \qquad (3\text{-}2)$$

Hence, for an *AT* cut to operate on 5 MHz, the thickness is 0.0132 in.:

$$t = \frac{k}{f} = \frac{66.2}{5000} = 0.0132$$

With a *BT* cut for the same frequency, however, the crystal would be thicker:

$$t = \frac{k}{f} = \frac{100.78}{50000} = 0.020156 \text{ in.}$$

When a sharp burst of electric energy is applied to a crystal slab (shock excitation), the oscillations which are produced decay more slowly than is the case with a resonant circuit composed of inductance and capacitance. Hence, the crystal has a much higher Q, providing for increased selectivity and efficiency in circuitry. The crystal's mechanical resonance can be compared to a series resonant circuit as shown in Fig. 3-3. The inductance L is comparable

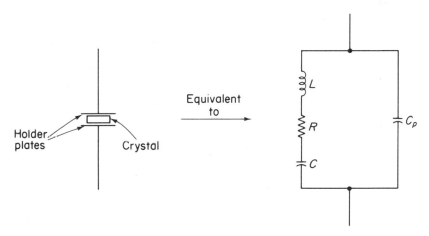

figure 3-3 crystal characteristics

to the crystal's mass, C to the compliance of the crystal, and R to the mechanical damping. Capacitance C_p is that of the metal holder plates. This capacitance is, of course, present whether or not the crystal is oscillating.

Although the crystal has a fundamental resonant frequency, a variable capacitor placed across it can tune above and below the fundamental frequency within limits depending on the type of cut, characteristics, etc. Attempts to tune too far from the fundamental frequency, however, will cause the crystal to stop oscillating. Increasing pressure of the crystal holder plates may also affect the frequency of crystal oscillations, again within limits, since too much pressure will crack the crystal.

Since the crystal is equivalent to a series-resonant circuit, Q is a function of the standard relation

$$Q = \frac{X_L}{R} \tag{3-3}$$

Thus, if a crystal is resonant at 600 kHz, its equivalent inductance (L) is 2.5 henrys (H), capacitance is 0.022 picofarads (pF), and equivalent resistance is 10,000 ohms (Ω), the Q is found to be 942:

$$Q = \frac{2\pi f L}{R} = \frac{6.28 \cdot 600,000 \cdot 2.5}{10,000} = 942$$

(The capacitance is not a function here since it is assumed that C has no internal resistance, and that only the inductance would have a resistive component.)

3-3. crystal oscillators

A typical transistorized crystal oscillator is shown in Fig. 3-4. An *npn* transistor is shown, though a *pnp* type can slso be used with a change in battery polarity. This is a grounded-base circuit [note the radio-frequency choke (RFC) in the emitter circuit that prevents the emitter from being at signal ground]. No radio-frequency choke is present in the base side, thus placing the base at signal ground by virtue of the low reactance of capacitor C_1.

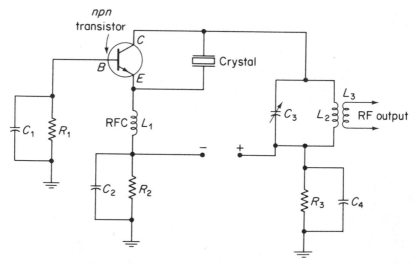

figure 3-4 transistorized crystal oscillator

With the grounded-base circuit there is no phase reversal between the signal at the collector and that at the emitter. Since signal currents in the emitter are in phase with those in the collector, the only requirement for oscillations is to provide coupling between collector and emitter. This coupling is provided by the crystal that forms a signal feedback loop and establishes the rate of oscillations.

The resonant circuit composed of L_2 and C_3 is part of the output system, with L_2 and L_3 forming a transformer. The necessary forward- and reverse-bias polarities are procured from a single battery or power-supply source, with resistors R_2 and R_3 forming a voltage-divider network across it. Thus, a ground connection is positive with respect to the emitter, but negative with respect to the collector.

A crystal oscillator using a triode tube is shown in Fig. 3-5. The crystal

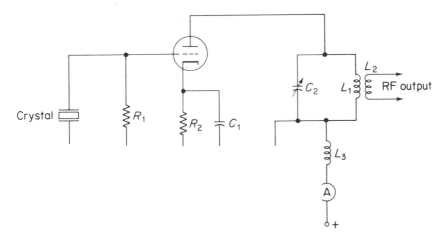

figure 3-5 tube-type crystal oscillator

forms a resonant circuit between the grid and cathode, as shown, while a parallel-resonant circuit composed of C_2 and L_1 is in the plate circuit. Resistor R_2 furnishes grid bias by making the cathode more positive than the grid (thus making the grid more negative with respect to the cathode). Capacitor C_1 acts as a bypass for signal currents, thus stabilizing the bias potential. This bias network also protects the circuit from excessive currents if the crystal fails to oscillate. Without R_2 the capacitance in the crystal and holder in conjunction with R_1 would establish grid bias, but failure to oscillate would cause a bias loss and consequent high plate currents.

Radio-frequency choke L_3 prevents signal energy from leaking to the power-supply systems, while C_3 provides the necessary signal return from the plate resonant circuit (also called the *tank* circuit on occasion) to ground, and hence to the cathode.

From basic electronics we know that an oscillatory circuit is formed by a combination of inductance and capacitance because of the energy exchange that occurs. If resistive losses are low, oscillations are sustained for a brief interval after shock excitation. A crystal or other resonant circuit (L and C) has only a resistive component when tuned to resonance for a specific frequency, since the capacitive reactance X_C is equal in ohmic value but opposite in polarity to the inductance reactance X_L; hence, the two reactances are essentially canceled.

For Fig. 3-5, if the resonant circuit in the anode is tuned *exactly* to resonance (the same frequency as the crystal), both the crystal and the anode-resonant circuit will be resistive, since reactances cancel. Now the *entire* oscillator circuit (in terms of lumped inductances and capacitances) would be unable to oscillate, since the required L and C effects are missing for any energy-exchange condition. Some extra capacitances may be present (tube or transistor internal capacities, plus stray capacity of the circuit wiring), but inductance is needed for circuit oscillations. Hence, it is necessary to tune the anode-resonant circuit slightly higher in frequency than the resonant frequency of the quartz crystal.

When a resonant circuit is tuned above resonance by decreasing the value of inductance or capacitance (or both), the lower-frequency oscillations that prevail in the circuit increase the capacitive *reactance* X_C and lower the inductive *reactance* X_L. Since the inductive reactance decreases, there is an increase in signal current flow though the inductance, causing an inductive characteristic. This inductance, in conjunction with lumped circuit capacitance, will now permit the complete oscillatory circuit to function properly.

For stability in operation, therefore, it is preferable to tune the output resonant circuit between the arrows, as shown in Fig. 3-6, and slightly above the normal dip in transistor collector or tube plate current, which occurs near the center frequency of oscillations. This procedure produces stability for the *desired* signal frequency oscillations.

At resonance the inductive reactance is equal but opposite to the

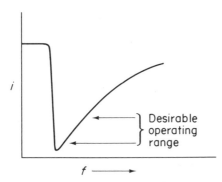

figure 3-6 preferred operating region of crystal oscillator

capacitive reactance; hence

$$X_L - X_C = 0$$

Showing the angular velocity (ω) instead of 6.28f, we obtain

$$\omega L - \frac{1}{\omega C} = 0$$

$$\omega^2 = \frac{1}{LC} \left(6.28f^2 = \frac{1}{LC}\right)$$

$$LC = \frac{1}{\omega^2}$$

$$f_r = \frac{1}{6.28\sqrt{LC}} \qquad (3\text{-}4)$$

Thus, Eq. (3-4) solves for the resonant frequency (f_r) when the inductance in henrys and the capacitance in farads are known. The following equation simplifies calcuations:

$$f_r = \frac{159}{\sqrt{LC}} \qquad (3\text{-}5)$$

where f_r = frequency of resonance in kilohertz
   L = inductance in microhenrys (μH)
   C = capacitance in microfarads (μF)

Thus, if the inductance for a given resonant circuit is 124 μH, and the capacitance is 0.0004 μF, the product of L times $C = 0.0496$; hence

$$f_r = \frac{159}{\sqrt{0.0496}} = 715 \text{ kHz}$$

and from Eq. (1-8), $\lambda = 420$ m.

On occasion it may be necessary to calculate the amount of inductance or capacitance necessary to tune to a specific frequency. To obtain resonance for a specific frequency, a particular inductor could have various values, depending on the capacitance present. Similarly, the capacitor value depends on the inductor value also, since the *product* of LC determines the resonant frequency (the L-to-C ratio). Thus, if LC is 0.0000045, the frequency of resonance is 75,000 kHz (4 m), and C or L can have any value so long as this product is obtained. The selection of a particular L or C value, however, affects circuit Q and bandwidth, as more fully discussed in Chapter 4.

The following formulas are useful for finding particular microvalues of L or C when one quantity is fixed and the frequency is known:

$$L = \frac{25,300}{f_r^2 C} \qquad (3\text{-}6)$$

$$C = \frac{25,300}{f_r^2 L} \qquad (3\text{-}7)$$

Thus, if the capacitor in a parallel-resonant circuit has a value of 0.002 μF and the desired frequency is 387 kHz, the inductance needed is

$$\frac{25,3000}{387^2 \times 0.002} = \frac{25,300}{299.538} = 84.5 \ \mu H$$

Since Eqs. (3-6) and (3-7) are derived from Eq. (3-5), we use microhenrys for the inductance, kilohertz for the frequency, and microfarads for the capacitance. Hence, the 84.5 obtained for the inductance represents microhenrys.

Similarly, assume that the inductance is known to be 50 μH and the desired frequency, 2000 kHz (2 MHz). What is the capacitance value?

$$C = \frac{25,300}{2000^2 \times 50} = \frac{25,300}{200,000,000} = 0.0001265 \ \mu F$$

Had the inductance been 25 μH, the following changes occur:

$$C = \frac{25,300}{2000^2 \times 25} = \frac{25,300}{100,000,000} = 0.0002530 \ \mu F$$

Thus, since the inductance value was halved, the capacitance value doubled to obtain resonance at the same frequency. Note that the same product of LC prevails:

$$50 \times 1265 = 63,250 \ (LC)$$

$$25 \times 2530 = 63,250 \ (LC)$$

Another solid-state crystal oscillator is shown in Fig. 3-7, in which use is made of the MOSFET (*M*etal *O*xide *S*emiconductor *F*ield *E*ffect *T*ransistor). The MOSFET combines the advantages of the transistor with that of the vacuum tube. As in transistors, the MOSFET is mechanically rugged, requires no filament potentials, and is small in size compared to vacuum tubes in equivalent use. As with a pentode tube, the MOSFET has linear

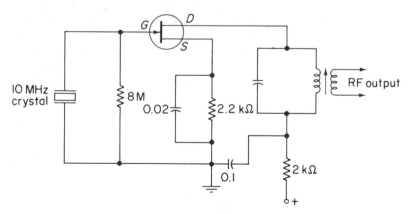

figure 3-7 *N*-channel MOSFET crystal oscillator

transfer characteristics and a high input impedance. In addition, it has low noise, less stray capacitance then transistors, improved thermal stability, and a much wider frequency range.

An *n*-channel MOSFET crystal oscillator is shown in Fig. 3-7, and the circuit resembles the tube-type oscillator shown earlier in Fig. 3-5. As with *npn* transistors or vacuum tubes, the *n*-channel MOSFET has a positive polarity applied to the output element. The *gate* (*G*) can be compared to the grid of a tube or the base of the junction transistors. The *source* (*S*) is similar to the emitter or cathode, and the *drain* (*D*) to the collector or anode. A *p*-channel MOSFET could be used, with reversal of battery or supply potentials.

For the oscillator shown, the output resonant circuit is tuned with a movable core (slug) instead of a variable capacitor, as shown for the oscillator in Fig. 3-5. Either method could be used for either oscillator.

3-4.   variable-frequency RF oscillators

There are occasions when it is necessary to employ a variable-frequency RF oscillator so that it can be tuned to a specific frequency as needed. One such application is the *local* oscillator in superheterodyne receivers, as shown in Fig. 3-8. Here the variable capacitors of the mixer and oscillator (as well as the RF stage if present) are ganged together, as shown by the dashed line. As the mixer resonant circuit is tuned for various stations, the oscillator frequency is also varied so that the frequency difference between the two

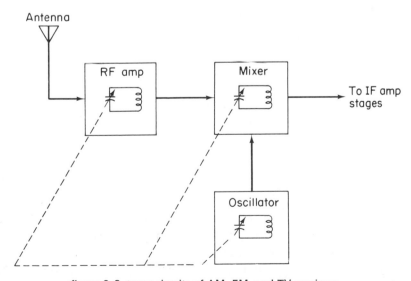

figure 3-8 tuner circuits of AM, FM, and TV receivers

circuits is maintained. This produces a single-frequency output (called *intermediate frequency*) (see Chapters 6 and 7). Other applications of variable-frequency oscillators are in test equipment and in some mobile transmissions when operation is in a band of frequencies rather than on a single assigned frequency.

A widely used tunable oscillator is the Hartley, shown in Fig. 3-9. The

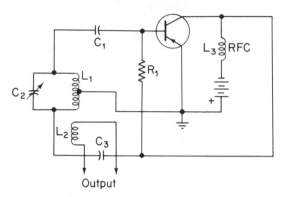

figure 3-9 Hartley oscillator

resonant circuit composed of C_2 and L_1 is shared by both the base and collector circuits because of the tap on the inductance. The lower portion of L_1 is across emitter and collector, since the tap is at ground (the same as the emitter), and the bottom of the coil is coupled to the collector using capacitor C_3. This capacitor prevents the shorting of the collector supply potential to ground through the tapped coil.

The base is coupled to the top of the inductance, thus forming the base-emitter resonant circuit. Hence, the collector signals are effectively coupled back to the base system to sustain oscillations. Resistor R_1 supplies the necessary forward bias to the base emitter (positive for the emitter and negative for the base), and the *RFC* provides a high inductive reactance for signal energy, thus preventing losses through the power-supply system. An *npn* transistor could, of course, be used also, with reversal of battery potentials. Vacuum-tube Hartley oscillators are similar, with the tapped coil placing part of the resonant circuit between grid and cathode and the other between cathode and anode. The output is obtained by L_2 coupled to the collector end of inductance L_1, as shown.

Other variable-frequency oscillators (VFO's) are shown in Fig. 3-10. The *Colpitts* oscillator (Fig. 3-10a) uses two variable capacitors across the inductance in the resonant circuit. Instead of a tapped inductance, the tap is at the junction of the two variable capacitors, as shown. Except for this factor, operation is similar to that of the Hartley. For the Colpitts shown, the collector is applied directly to the resonant circuit, with the variable capaci-

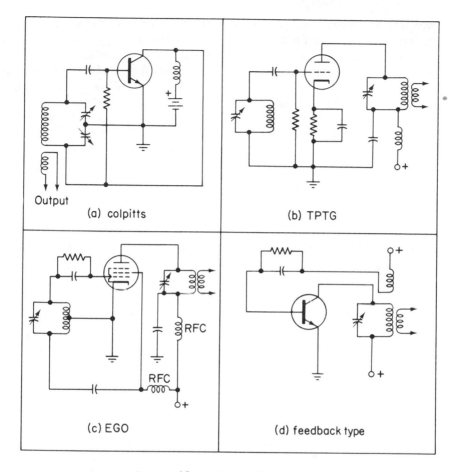

Output

(a) colpitts

(b) TPTG

(c) EGO

RFC

RFC

(d) feedback type

figure 3-10 additional RF oscillators

tors providing dc isolation to ground. The base capacitor is necessary to prevent direct coupling between lase and collector.

The *tuned-plate tuned-grid* (TPTG) oscillator is shown in Fig. 3-10b. A transistor in place of the triode could also be used, with a tuned circuit between base and emitter as well as between collector emitter. The operation of this type of oscillator is similar to the crystal type, except that it is tunable over the range permitted by the variable capacitor-coil combinations. As with the crystal oscillator, the output resonant circuit must be tuned slightly higher in frequency than that which the oscillator is to generate.

The electron-coupled oscillator (ECO) shown in Fig. 3-10c has a frequency stability superior to that of the ordinary variable-frequency oscillator, and with careful circuit design its ability to maintain a given frequency output is surpassed only by the crystal-oscillator type. As shown, a pentode tube is used and the resonant circuit at the grid side is shared by the screen grid acting as an anode for the Hartley-oscillator section. Since the screen grid acts as an anode, the circuit oscillations are present in the electron stream flowing from cathode to screen grid. Anode current for the pentode, however, also flows between the cathode and anode; hence, the signal current variations that occur between cathode and screen grid influence the plate current flow. Thus, the resonant circuit in the anode section of the ECO is pulsed in a timing sequence coinciding with the frequency of the Hartley-oscillator section of the system. Hence, the anode resonant circuit is *electron coupled* to the oscillator section, and loading effects normally found in direct (or inductive) coupling are kept at a minimum. Consequently, load applications to the output resonant circuit are effectively isolated from the true oscillator section; thus, stability is increased considerably.

One of the earliest oscillator circuits is the *feedback* type shown in Fig. 3-10d. A feedback inductance, sometimes called a *tickler coil*, couples some of the oscillatory signal energy developed in the collector circuit back to the base circuit to sustain oscillations. The feedback coil must be wound in the direction that coincides with that of the collector inductance for proper feedback polarity. The degree of coupling is adjusted for maximum amplitude consistent with the best possible stability.

3-5.   blocking oscillator

As mentioned in the Section 3-1, *relaxation* oscillators do not employ resonant circuits to sustain oscillations. Instead, combinations of capacitors and resistors are used to form charge and discharge circuits for alternately blocking electron flow through the transistors or tubes that make up the system.

There are two basic types of relaxation oscillators: the *blocking oscillator* and the *multivibrator*. Typical use in television receivers is shown in Fig. 3-11. Here the vertical- and horizontal-sweep signals are developed by relaxation oscillators of either the blocking oscillator or the multivibrator type.

Transistor and tube-type blocking oscillators are shown in Fig. 3-12. For the *transistor* type (Fig. 3-12a) an *npn* type could be used instead of the *pnp* transistor, with reversal of battery potential. Since this oscillator uses feedback for oscillation purposes, it is necessary to invert the phase of the signal appearing at the collector so that it will be in phase with the signal appearing at the base. This is done by the transformer shown in which the secondary terminal connections are made for proper phasing of the feedback signal.

The high degree of feedback causes the base of the transistor to be driven beyond the cutoff region for a time dependent on the time constant of the

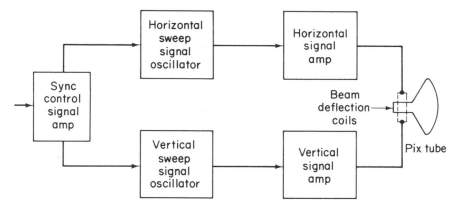

figure 3-11 relaxation oscillators in television

circuit. A charge is developed across the capacitor in the base circuit, which opposes the normal reverse bias (negative with respect to the emitter for a *pnp* transistor). This charge gradually leaks off and conduction again occurs, repeating the cycle.

As shown in Fig. 3-12a, the waveforms are not sinusoidal and, hence, are rich in harmonics. The output waveforms are applied to waveshaping circuits (discussed later) to form the required square waves, sawtooth signals, or other types. The frequency of the signals generated depends on the capacitance and resistance values of the circuit, including the internal characteristics of the transistor and those of the transformer; hence, the exact frequency is difficult to predict mathematically. The frequency of the output signals can be altered by adjustment of the variable resistor in the base circuit, by changing the capacitance of the base capacitor, and by shunting the L_1 (primary) section of the transformer with capacitors. If oscillations are not produced, L_2 terminals may have to be reversed.

The tube-type blocking oscillator shown in Fig. 3-12b functions similarly to the transistor type. The high degree of feedback alternately blocks the grid by charging capacitor C_1 so that a negative potential appears at the grid and drives it beyond the cutoff point. Conduction stops and the capacitor discharges through resistor R_1 until the bias on the grid is reduced sufficiently to permit conduction again, at which time the cycle is repeated.

Instead of using the feedback winding for output purposes, a separate winding (called a *tertiary*) is shown in Fig. 3-12b. This separate output winding minimizes loading effects to some extent, hence reducing frequency changes introduced by the load characteristics. Such a tertiary winding could also be used for the transistor circuit shown in Fig. 3-12a.

Synchronization for the oscillator in Fig. 3-12b is introduced into the grid circuit by resistor R_2. This synchronization method is also applicable to the transistor circuit. Since such oscillators (including the multivibrators

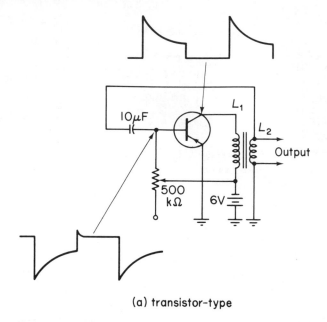

(a) transistor-type

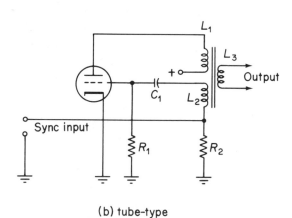

(b) tube-type

figure 3-12 blocking oscillators

discussed later) undergo periodic blocking of conduction, the introduction of a pulse slightly earlier than the normal start of each conduction period can control the frequency.

The synchronization principle is illustrated in Fig. 3-13. The signal (Fig. 3-13a) represents the waveform at the grid of a tube or base of a transistor. For a tube, the waveform represents a climb into the conduction region from a high negative value. Thus, the grid becomes less negative until the bias is low enough to permit conduction. The signal may climb above the

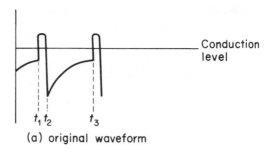

Conduction
level

$t_1\,t_2$ t_3

(a) original waveform

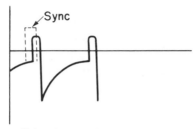

Sync

(b) pulse position

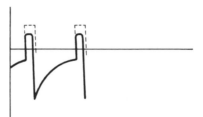

figure 3-13 synchronization in
relaxation oscilla-
tors   (c) synchronized waveform

conduction region and become slightly positive before it plunges into the cut-off region again, as shown. For an *npn* transistor the signal represents a change from a negative potential (reverse bias) to a positive potential (forward bias) that permits conduction, with opposite polarity conditions for the *pnp* transistor.

If a pulse is applied to the input circuit slightly earlier than normal (Fig. 3-13b), conduction will occur sooner and the free-running signal frequency will now be synchronized with the input signal, as shown in Fig. 3-13c, since the incoming pulses now control the time when the oscillator transistor (or tube) goes into conduction. Thus, if the frequency of the incoming pulse (repetition rate) is sufficiently close to the oscillator frequency, it will synchronize the oscillator and control its frequency. The sync pulse must be *close* to t_1 or t_3 in Fig. 3-13a. If the sync pulse appears *between* t_2 and t_3, its amplitude may not be sufficient to reach the conduction level. (particularly if the pulse appears near the bottom of the waveform, near t_2)

For the system shown in Fig. 3-11, the sync signals are transmitted by the television station and thus lock in the sweep circuits by remote control. In this manner, vertical and horizontal sweep of the picture signal is synchronized with that at the station to which the receiver is tuned, even though many miles away.

3-6. multivibrator

The multivibrator is another type of relaxation oscillator that depends on intermittent conduction for sustaining oscillations. Instead of a transformer, another transistor (or tube) is used for phase inversion and feedback purposes. Also, as with the blocking oscillator, the multivibrator can be synchronized by an external signal for frequency-control purposes.

A typical multivibrator is shown in Fig. 3-14, where two *npn* transistors

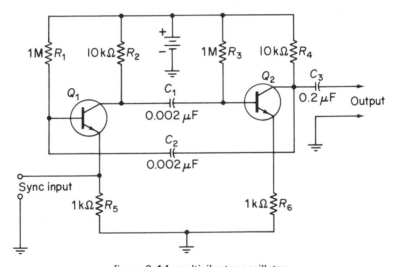

figure 3-14 multivibrator oscillator

are used. *Pnp* types could, of course, also be employed with appropriate polarity changes of the source voltage. Note that the output signal at each collector is coupled to the base of the other transistor. Since the common-emitter circuitry provides for a phase reversal of the signal from base to collector, the output from Q_1 is out of phase with respect to the input. This out-of-phase signal is coupled to the base of Q_2, and at the collector the phase is again reversed, bringing it back to the original at the input of Q_1. Thus, by coupling the output of Q_2 back to the input of Q_1 (via capacitor C_2) we have an in-phase feedback loop to form an oscillatory system.

The circuit shown in Fig. 3-14 is symmetrical, having identical-value resistors in each base circuit and collector circuit, as well as similar-value capacitors and transistors with the same characteristics. Despite this, however, when power is first applied, current flow will rise more rapidly for one transistor than the other (a necessity for oscillations to start).

Assume, therefore, that Q_1 initially conducts more current than Q_2. During conduction of Q_1, the voltage drop across collector resistor R_2 charges capacitor C_1 so that it has a negative polarity at the base of Q_2 and a positive polarity toward the Q_1 collector. As the negative potential at the base of Q_2 increases, conduction through this transistor decreases because of the lowered forward bias (for the *npn* a positive bias is required at the base with respect to the emitter). As conduction through Q_2 decreases, the positive potential at the collector of Q_2 will increase. This increase is felt at the base of Q_1 because of the coupling capacitor C_2. The increase in the forward bias at the base of Q_1 increases conduction through this transistor. Conduction continues until saturation is reached, at which time transistor Q_2 is cut off and no longer is there a voltage or current *change*. Since the *change* no longer occurs, capacitor C_1 now discharges and, within a short time, permits Q_2 to conduct again. With Q_2 conducting, its collector potential decreases, and this change of potential charges C_2 with a negative polarity toward the base of Q_1. This has the effect of reducing the forward bias of Q_1 with the resultant decrease in conduction. The process continues rapidly until Q_1 is cut off and Q_2 conducts at saturation. This now represents the reverse of the original condition and the process starts over again, thus sustaining oscillations.

Since a symmetrical circuit is involved, the sync input could be across R_6 instead of R_5. Similarly, the output could be obtained from the collector of Q_1 instead of at Q_2. The frequency of operation depends on values of R and C, as well as on the internal characteristics of the particular transistors used.

3-7. *sawtooth formation*

The sawtooth waveform necessary to sweep a beam across the face of a picture tube (or oscilloscope tube) can be obtained from the relaxation oscillators by the addition of discharge circuitry, which gradually charges a capacitor and discharges it rapidly. A typical discharge system for sawtooth formation is shown in Fig. 3-15 and uses a transistor for precise control of the discharge interval.

As shown, no forward bias is applied between the base and emitter; hence, the transistor is cut off. During normal operation, however, a signal

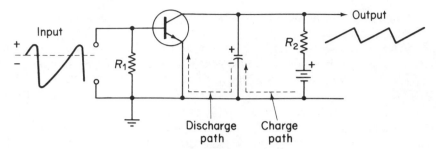

figure 3-15 discharge system sawtooth formation

is applied constantly to the discharge system from the output of a blocking oscillator or multivibrator. Thus, a varying potential exists at the input of the discharge circuit, which holds the base emitter at cut off for a given period and permits conduction for a brief interval when the input signal has a positive polarity (and thus supplies the necessary forward bias for the *npn* transistor).

During the time the transistor is in the nonconduction state, the capacitor charges, with electron flow in the direction shown from the negative battery terminal. The rate and amplitude of charge depend on the values of the capacitor, the resistor, and the battery potential. As the capacitor charges, it produces a gradually rising voltage between collector and ground, which forms the initial incline of the sawtooth signal being developed. The inclination is virtually linear up to approximately 10 per cent of the full charge of the capacitor, and for a good sawtooth waveform it is desirable to discharge the capacitor during the time interval prior to this 10 per cent charge (a time constant RC of 0.1, as shown in Table 3-1). If the capacitor is discharged too far beyond this point, sawtooth distortion results and the desired straight-line incline will take on an exponential curve.

When the input waveform reaches a positive value sufficient to cause transistor conduction, the transistor's low internal impedance shunts the capacitor and permits it to discharge through the transistor in the direction shown in Fig. 3-15. Thus, since the transistor now conducts, electrons leave the negatively charged side of the capacitor and flow through the transistor from emitter to collector and to the positively charged side of the battery, thus providing for a sudden discharge and forming the decline portion of the sawtooth waveform.

When the incoming signal again becomes negative, the transistor stops conducting and the capacitor starts to charge again, repeating the entire process and forming an output train of sawtooth waveforms, the frequency of which is controlled by the applied input signal. During the charging interval the time constant is long compared to the discharge, since the charge path is through the resistor. During discharge, however, the time constant is short because the path is through the low internal resistance of the transistor during conduction.

table 3-1   time-constant values

time constant	percentage of capacitor discharge voltage or charge current	percentage of capacitor charge voltage
0.001	99.9	0.1
0.002	99.8	0.2
0.003	99.7	0.3
0.004	99.6	0.4
0.005	99.5	0.5
0.006	99.4	0.6
0.007	99.3	0.7
0.008	99.2	0.8
0.009	99.1	0.9
0.01	99	1
0.02	98	2
0.03	97	3
0.04	96	4
0.05	95	5
0.06	94	6
0.07	93	7
0.08	92	8
0.09	91	9
0.10	90	10
0.15	86	14
0.20	82	18
0.25	78	22
0.30	74	26
0.35	70	30
0.40	67	33
0.45	64	36
0.5	61	39
0.6	55	45
0.7	50	50
0.8	45	55
0.9	40	60
1	37	63
2	14	86
3	5	95
4	2	98
5	0.7	99.3

The charging characteristics of the system are regulated by the value of R_2, which controls the amplitude of the sawtooth signal formed. In vertical sweep systems of test equipment, television receivers, and other systems, R_2 is made variable, or an additional resistor is placed in series with R_2 for amplitude control. The charge and discharge factors have the following relationships:

$$\tau = RC \tag{3-8}$$

$$C = \frac{\tau}{R} \tag{3-9}$$

$$R = \frac{\tau}{C} \tag{3-10}$$

$$e_C = E(1 - \epsilon^{-t/RC}) \tag{3-11}$$

where τ = time constant in seconds
R = total series resistance in ohms
C = capacitor value in farads
e_C = instantaneous voltage across the charging capacitor
E = maximum voltage (source voltage)
ϵ = Naperian log base (equals 2.718)
t = time in seconds

Equations (3-8), (3-9), and (3-10) are standard time-constant formulas. Thus, if the capacitance is 0.2 μF and the resistor in series with it is 10,000 Ω one time constant is 0.002 s, indicating the time at which the capacitor charges to 63 per cent of full value:

$$\tau = RC = 0.2 \times 10^{-6} \times 10,000 = 0.002$$

If the resistance and time constant are known, Eq. (3-9) is useful for calculating the required capacitance. As an example, assume that the capacitor in a discharge circuit must charge to 10 per cent of the applied voltage in 200 microseconds (μs). If the resistance is 10,000 Ω, what must be the value of the capacitance?

From Table 3-1 we find that the 10 per cent capacitor charge represents a time constant of 0.1. If 0.1τ must equal 200 μs (0.2 ms), one complete time constant must equal

$$\tau = \frac{200}{0.1} = 2000 \ \mu s$$

$$C = \frac{\tau}{R} = \frac{2000 \times 10^{-6}}{10,000} = 0.2 \ \mu F$$

When the capacitance and time constant are known but the resistance must be found, we use Eq. (3-10). Thus, assume that a capacitor must charge to 8 per cent of full value in 100 μs. If the capacitor has a value of 0.005 μF, what must the resistance be?

From Table 3-1 the time constant is shown to be 0.08 when the capacitor has a charge of 8 per cent of full value. If $0.08 \ \tau$ must be 100 μs, then one time constant (1 τ) is 100/0.08 and is, therefore, 1250 μs. Applying this to Eq.

(3-10), we obtain

$$R = \frac{1250}{0.005} = 250,000 \ \Omega$$

Equation (3-11) is useful for precise calculations of the instantaneous voltage across the charging capacitor in an RC circuit and eliminates the need for using Table 3-1. The table is not perfectly accurate, since percentages of voltages and currents are taken to the nearest two-digit value based on values obtained from tables of exponential functions. A τ of 0.7, for instance, gives a percentage value of capacitor charge voltage of 49.66 instead of the 50 shown in the table. Also, the table does not show intermediate time constants such as 0.063, 0.22, etc.

When using Eq. (3-11), it is necessary to refer to the table of exponential functions (Appendix I) for determining values of ϵ^{-x}, where ϵ is the Naperian log base previously mentioned, and x is t/RC. To show the application of Eq. (3-11) as well as to compare the results with calculations using Table 3-1, two examples are given.

Assume that a 150,000-Ω resistor is in series with a 0.0005-μF capacitor and a 12-V source. After 150 μs what is the instantaneous voltage across the capacitor?

Solution: One $\tau = RC = 150,000 \times 0.0005 \times 10^{-6}$

$$= 75 \ \mu s$$

Hence, 150 μs $= 2 \ \tau = 86$ per cent of full charge.

Answer: $e_C = 12 \times 0.86 = 10.32$ V

Using Eq. (3-11),

$$e_C = 12(1 - \epsilon^{-x})$$

Solving for x,

$$x = \frac{t}{RC} = \frac{150}{75} = 2$$

Using the exponential function table (Appendix I), we find that for $-x = 2$ we obtain a value of $\epsilon^{-x} = 0.135$. Subtracting this from 1 and multiplying by 12 provides the answer:

$$12 \times (1 - 0.135) = 12 \times 0.865 = 10.38 \text{ V}$$

3-8.   pulse characteristics

As discussed in Chapter 1, the square wave can be compared to the sinewave, since both have positive and negative alternations and, hence, ac characteristics. Pulses, however, consist of only single-polarity signals and thus are either positive or negative, with differences as shown in Fig. 3-16.

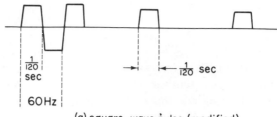

(a) square-wave pulse (modified)

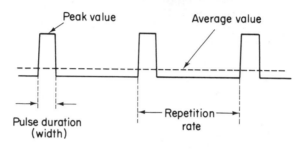

(b) short duration pulses

figure 3-16 pulse factors

Figure 3-16a shows a square wave with a frequency of 60 Hz, with each alternation of the square wave thus having a duration of $\frac{1}{120}$ s. If we eliminate the negative alternations of the square waves, we obtain pulses also having a duration of $\frac{1}{120}$ s.

Pulses, however, could still have a duration of $\frac{1}{120}$ s without having a frequency of 60 Hz, since the individual pulses could occur at a rate of 20/s, or 10/s, etc. The rate at which pulses occur per second therefore differs from the frequency of sinewaves and squares and is termed the *repetition rate*. Thus, a pulse may have a duration of $\frac{1}{45}$ s, or $\frac{1}{1000}$ s, but its recurrence may be at a very low rate, or, if required, at an extremely high rate. This is not the case with an unbroken chain of square waves or sinewaves, since such waveforms have a frequency fixed by the duration of the individual alternations comprising the signals.

Since the duration and repetition rate of a pulse train are no longer dependent on each other, as is the case with sinewaves, both factors must be considered to indicate the characteristics of pulses. The term *duty cycle* is a designation relating to both the repetition rate and duration of a pulse train:

$$\text{duty cycle} = \text{pulse duration} \times \text{repetition rate} \qquad (3\text{-}12)$$

Thus, if each pulse in a series has a duration of 3 μs and the pulses are repeated 500 times per second, the duty cycle would be 0.000003 times 500, or 0.0015.

With sinewaves the effective value of voltage or current is obtained by multiplying the peak value by 0.707. This procedure, however, is not applicable to a pulse train, since the repetition rate is a determining factor relative to the power present. Usually, the average power of a series of pulses is much lower than the 0.707 value prevailing in sinewaves. As shown in Fig. 3-16b, with the pulses widely spaced, average power drops to a low level.

Knowing the duty cycle, we can solve for the average power of a pulse train by

$$\text{average power} = \text{peak power} \times \text{duty cycle} \qquad (3\text{-}13)$$

Thus, if the duty cycle for a pulse train is 0.0015 and the peak power of a pulse is 10 watts (W), the average power would be $0.0015 \times 10 = 0.015\,\text{W}$. Obviously, if the average power and the peak power are known, we can solve for the duty cycle by dividing the average by the peak.

As described in Chapter 1, square waves and pulses are made up of a number of harmonic components, each higher harmonic having an amplitude lower than the previous harmonic. In contrast to square waves, the pulse waveforms contain both odd and even harmonics, and the more narrow the pulse, the greater the harmonic content. The maximum number of successive higher harmonic components that have a bearing on the fidelity of the pulse shape varies inversely with respect to pulse duration. Wide pulses may have significant harmonic components only to the tenth or fifteenth order, while narrow pulse waveforms may have a significant harmonic component range to 1000 or higher.

The rise time of the leading edge of a pulse represents a high frequency, as does the decline or decay time of the trailing edge. If circuitry is not well designed and limits high-frequency response, there will be distortion of the original waveform. A pulse without some of the higher-frequency signals will no longer have sharp changes in the leading and trailing edges, but rather the changes become more sloping.

There is pulse distortion also for loss of low-frequency components, though the effects on pulse shape are not as severe as the loss of higher-signal components. Some of the lower-frequency components of a pulse may, in fact, be removed entirely without seriously altering pulse shape. Thus, low-frequency response in a pulse amplifier is not as critical a necessity as high-frequency response.

An indication of the extent of high-frequency components that must be retained during amplification is given by the *base frequency:*

$$\text{base frequency} = \frac{1}{\text{pulse duration}} \qquad (3\text{-}14)$$

Equation (3-14) indicates the highest frequency component that a circuit must pass for good reproduction of the pulse during amplification. Thus, if each pulse in a series has a duration of 4 μs, the base frequency is 25 kHz and circuits through which such a pulse passes should be capable of handling signal frequencies from the repetition rate up to 25 kHz. If the repetition rate is 1500 Hz, this would set the lower limit of response deemed desirable.

Since the repetition rate is not related to the harmonic content of a pulse, it is not included in Eq. (3-14). The repetition rate simply indicates the number of times a pulse appears per second and has no effect on the harmonic content of individual pulses.

Amplifiers and other circuits handling pulse and square-wave signals are designed to have a maximum response range for low, intermediate, and high frequencies to minimize distortion. There are, however, many occasions when some of the characteristics of pulses are deliberately altered to meet specific requirements. Thus, some of the high-frequency components may be attenuated, or portions of the pulse may be clipped, etc., as described in the remainder of this chapter.

3-9. integrating circuits

When a shunt capacitance exists in the circuit of Fig. 3-17a, it can attenuate high-frequency signals or the high-frequency components of pulse-type waveforms. The degree of attenuation depends on the capacitive reactance $[1/(6.28 \cdot f \cdot C)]$ and, hence, on the frequency of the signals and the capacity present. Thus, when good circuit fidelity is required, such shunt

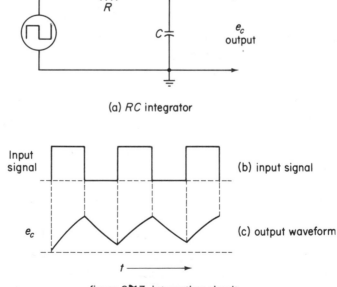

(a) RC integrator

(b) input signal

(c) output waveform

figure 3-17 integrating circuit

capacitances between wires, between solid-state elements, etc., are kept at a minimum by careful component selection and wiring design.

On occasion a shunt capacitance is deliberately introduced to form what is known as an *integrating* circuit. Such a system is useful in electronic counting systems and in the vertical-sweep sections of television receivers, as more fully discussed in Chapter 8. The basic integrating circuit includes the series resistor shown in Fig. 3-17a.

In practical applications, the integrating circuit can be considered equivalent to a low-pass filter. For sinusoidal signals there will be progressive attenuation of signals with higher frequencies, though no waveform modification will occur. For pulse or square-wave signals, however, the waveform is altered because of the attenuation of the higher-frequency harmonic components making up the waveform.

What occurs when potentials are applied to a capacitor should be considered for an understanding of integration. When the applied voltage is direct current, electrons flow to one capacitor plate and away from the other as the capacitor charges. For sinewave signals, however, the repetitive polarity reversals cause periodic charges and discharges of the capacitor in opposite directions, conforming to the frequency of the applied signal. Electron flow to and from the capacitor can be considered the *capacitor signal current*. From the calculus, the applied signal voltage e_C and the capacitor signal current i_C have the following relationship:

$$e_C = \frac{1}{C} \int i_C \, dt \qquad (3\text{-}15)$$

where e_C = signal voltage across the capacitor
C = capacitance in farads
i_C = capacitor signal current

Note that e_C is proportional to the time interval of i_C. In practical circuit integrators the time constant RC is *long* compared to the pulse width. Since a pulse contains a high order of harmonic signal components in addition to its fundamental frequency signal, we find that for the upper harmonic components for which the resistance value is much greater than the capacitive reactance of the capacitor, Eq. (3-16) applies:

$$e_C = \frac{1}{RC} \int e \, dt \qquad (3\text{-}16)$$

Thus, for an integrating circuit the output signal voltage is proportional to the *integral* of the input signal current. The results of integration are shown in Fig. 3-17b and c. Positive input pulses are shown, though integration occurs for pulses of either polarity. For the initial pulse, its steep leading edge represents a short time interval of applied voltage. The flat top of the pulse maintains the peak voltage across the circuit for a time equal to the duration of the pulse. For a capacitor, the voltage buildup lags the current, and voltage rises exponentially. During one time constant (RC) the capacitor would attain approximately 63 per cent of full charge, with approximately five time constants needed for a full charge. However, since the integrator has a long time constant compared to pulse width, capacitor charge voltage does not attain a maximum leveling-off value. Instead, there is a gradual incline (Fig. 3-17c).

When the trailing edge of the input pulse arrives, the signal-voltage amplitude drops to the zero level in a very short time interval. Now the capacitor discharges through the series resistor and the input circuit. The rate of discharge is again slow compared to the trailing-edge time interval and the result is an output signal waveform (Fig. 3-17c). This waveform indicates attenuation of higher-frequency components, and the actual waveshape depends on the precise time constant prevailing relative to the width of the input pulse.

The pulse train (Fig. 3-17b) has intervals between pulses the same as the pulse widths. If the intervals between pulses were long, it would permit the capacitor to discharge fully between pulses. If the time interval between pulses is much shorter, complete discharge does not occur. For short-duration pulses with shorter time between them, there would be a gradual voltage buildup across the integrator capacitor. This system is often used as a frequency divider, since the firing level of a relaxation oscillator is reached only after a designated number of input pulses have arrived to build up the charge to that required. This is discussed more fully in Chapter 8.

3-10.   differentiating circuits

A series capacitor, such as shown in Fig. 3-18a, is often used as a coupling device between two successive circuits. Since a capacitor has an increasing reactance for lower-frequency signals, such a series capacitor will attenuate low-frequency signals to the degree of reactance present. Thus, to minimize low-frequency losses, high-value capacitors are used for interstage coupling, or direct coupling is used (without an intervening capacitor or transformer).

As with the integrating circuit, losses for specific signal frequencies are often introduced deliberately to form a required pulse modification. With the series capacitor and shunt resistance as shown in Fig. 3-18a, and using a short time constant compared to the width of the applied pulse, a circuit is formed that is termed a *differentiating circuit*. It forms sharp, narrow pulses with steep leading edges that are ideal for precise triggering purposes. Thus, the differentiating circuit retains the leading edge of the input pulse in the same polarity relationships and, basically, acts as a high-pass filter by attenuating the low-frequency components of the input pulses.

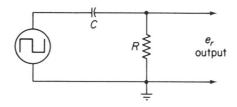

(a) *RC* differentiator

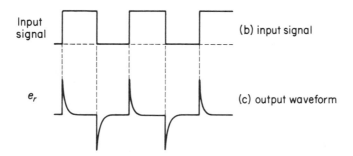

figure 3-18 differentiating circuit

For the circuit of Fig. 3-18, when voltage is applied the resultant current flow is proportional to the time derivative of the voltage appearing across the capacitor. This is shown by

$$i = C \frac{de_C}{dt} \tag{3-17}$$

Since the differentiator has a short time constant, some signal components of the input pulse will find that the resistor of the circuit has an ohmic value much lower than that of the reactance of the capacitor, since the resistor has a fixed ohmic value but the reactance of the capacitor depends on frequency and the value of capacitance. Hence, the voltage across the resistor R in Fig. 3-18a is

$$e_R = iR = RC \frac{de}{dt} \tag{3-18}$$

The input-signal waveforms are shown in Fig. 3-18b and the differentiated output in Fig. 3-18c. The leading edge of the input pulse represents a sharp increase in voltage applied to the input. In a capacitor, however, current leads voltage; hence, initially, current flow is high for the capacitor, and this flow through the output resistor produces the sharp leading edge of the differentiated output pulse waveform. The steady-state input voltage (flat top of pulse) offers no change; hence, the capacitor current starts to decline as the capacitor charges at a time rate dependent on the time constant of the circuit.

Since the time constant is short, the capacitor charges rapidly and current flow stops quickly. Hence, the voltage across the output resistor drops to zero as soon as the capacitor has become fully charged, and current flow ceases.

When the trailing edge of the input pulse arrives, the voltage applied to the input drops to zero, and the capacitor now discharges. The discharge direction is opposite to the charge direction and, in consequence, current flow through the resistor is now opposite to that of the charging current. Now a negative-spike signal is produced, as shown.

Differentiators and integrators can also be formed from R and L circuits, as shown in Fig. 3-19. Note that the differentiator (Fig. 3-19a) has

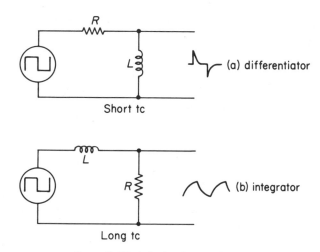

figure 3-19 L-R circuit waveforms

a series resistance and a shunt inductor, while the integrator (Fig. 3-19b) has a series inductor and a shunt resistor. The RC types are preferred, since the inherent internal resistance of an inductor could disturb circuit characteristics and make for difficult design.

3-11. clippers and clampers

Clipping and clamping circuits are widely used for removing noise or other undesired components from signal waveforms. Also, they are useful for clipping sinewaves to form square waves, for forming pulses from square waves, for maintaining constant amplitudes in signal waveforms, and for keeping a fixed dc reference level in certain circuitry.

The basic circuits for series and shunt clippers are shown in Fig. 3-20. The series clipper (Fig. 3-20a) has positive bias, with the output load resistor in series with the clipper diode. The 3-V bias has a polarity opposite to that

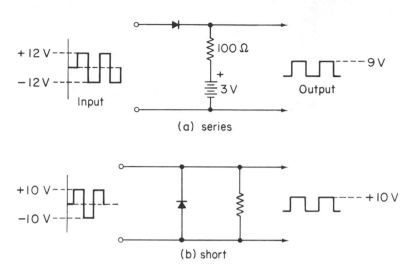

figure 3-20 clipper (series and shunt types)

which would cause diode conduction; hence, no current flows through the resistor until the positive polarity of an input signal exceeds the 3-V bias. For the negative alternations of the input signal, the diode does not conduct; hence, the circuit clips these portions of the input signal. Thus, the output signal has an amplitude which is proportional to that amplitude of the input signal that exceeds the bias potential.

A shunt clipper without bias is shown in Fig. 3-20b. For a positive alternation of the input signal, the diode does not conduct and, thus, presents a virtually infinite impedance across the output load resistor with no effect on the positive-polarity output signal. For negative alternations of the input signal, however, the diode conducts, and its forward resistance is now so low that it acts as an effective shunt across the output resistor. Consequently, no output signal is produced for a negative-polarity input signal. The result is a series of pulses, as shown.

A shunt clipper with positive bias is shown in Fig. 3-21a. Again the bias potential is opposite to that which would permit diode conduction and can be set at any desired value required for a particular degree of clipping. For the circuit shown, diode conduction only occurs when the positive amplitude of the input signal exceeds 3 V. The effect on several signal types is shown below the drawing.

If the input signal is a square wave (Fig. 3-21b) of 6 V (12 V peak-to-peak), an output signal amplitude in proportion to the input will develop up to a 3-V maximum in the positive direction. Beyond that amplitude the diode conducts and shunts the signal around the output resistor. For negative alternations of the square wave, however, the diode is nonconducting and the output is proportional to the input. Similarly, if the diode and the bias

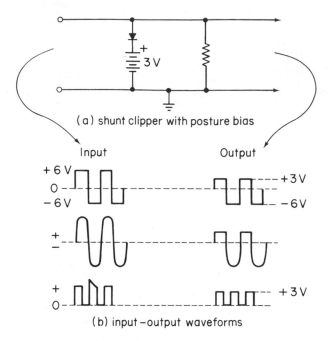

(a) shunt clipper with posture bias

(b) input–output waveforms

figure 3-21 clipper waveforms

potential were reversed, the negative portions of the output signal would be clipped.

The clipper of Fig. 3-21 can also be used for clipping either the negative or positive alternations of sinewaves, depending on the polarity of the bias battery. For that shown, the result is a flat-top production of the positive output alternations, as shown below the drawing.

The shunt clipper is also useful for limiting pulse amplitudes, as shown in the lower waveform drawings in Fig. 3-21b. With a positive-polarity bias of 3 V, the output amplitudes will be held at this constant 3-V amplitude because of the shunt effect created by the diode when input signal amplitudes are sufficient to cause conduction and shunting. Thus, if the input pulse train has a varying amplitude for the various pulses, the output will be held at a constant amplitude, as shown. This is also useful for removing transients or other undesirable distortion from the top of pulses.

As shown in Fig. 3-22a, two diodes can be used (with opposite-polarity wiring) to form a parallel clipper (also referred to as a *slicer*). The dual diodes permit the clipping of both the positve and negative alternations of a sinewave with the resultant formation of a square wave at the output, as shown. As with other clippers, the degree of waveform slicing can be regulated by setting the bias potentials with respect to the amplitude of the input signal. An overdriven amplifier (Fig. 3-22b) can also be used to form square waves from

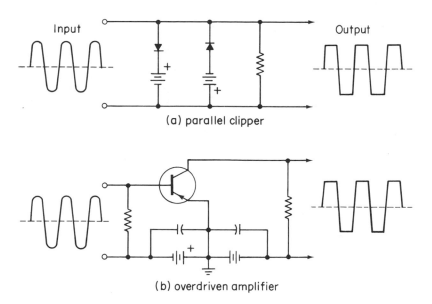

(a) parallel clipper

(b) overdriven amplifier

figure 3-22 parallel clipper functions

sinewaves by using a higher than normal amplitude input signal, driving the transistor into saturation.

Clamping circuits are used to restore the dc component of a pulse train when capacity coupling is employed in amplifiers or other circuits through which pulse signals must pass. The loss of the dc component through a capacity-coupled stage is shown in Fig. 3-23a. The amplified pulse train appearing at the collector of transistor Q_1 has a positive polarity, as shown. Assume that there is a 2.5-V drop across the collector load resistor R_L during signal absence. When the signal appears, the voltage *change* causes a rise to 7.5 V for the peak of the pulse; hence, the pulse amplitude is 5 V. The dc component is present and is equal to the *average* voltage of the pulse train.

When a capacitor is used for coupling purposes between transistor stages (or tubes), the dc level is lost, because the coupling capacitor cannot pass dc signal components. Consequently, the pulse train now applied to the input of the second transistor stage has a zero reference level with positive and negative alternations so that it is essentially a square wave instead of a series of single-polarity pulses. To revert back to the original waveform, it is necessary to clamp the waveform to a dc or zero level to restore the initial signal characteristics. The clamping process is often termed *dc restoration*.

For clamping purposes a diode is added in shunt to the R_1 resistor at the base of Q_2, as shown in Fig. 3-23b. At time t_1, during which no input signal appears, the capacitor C charges through the collector load resistor R_L with a polarity and direction as shown in Fig. 3-23a. At time t_2 the input pulse

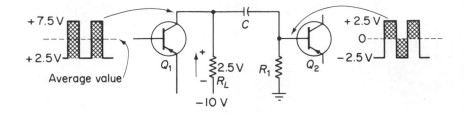

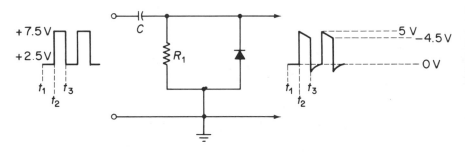

(a) capacity coupling between transistor stages

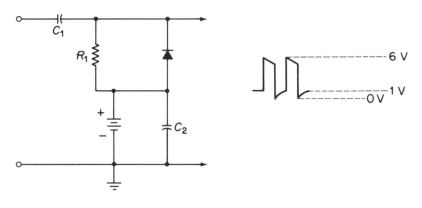

(b) clamping diode for dc restoration; modification to circuit shown in (a) above

(c) Circuit modification including dc bias

figure 3-23 clamping (dc restoration)

signal appears and voltage rises to 7.5. Because of the sharp rise time of the leading edge of the pulse, the capacitor does not have time to charge appreciably. The high leading current during this brief interval places the full pulse amplitude across the resistor. Hence, the potential across the resistor is the difference between the 7.5-V peak of the pulse and the idling voltage charge across the capacitor. This is so because the 7.5-V peak pulse amplitude makes the polarity of the input terminals to the clamper opposite to the capacitor charge, or 5 V. Thus, the output waveform rises to this 5-V peak value, as shown in Fig. 3-23a.

The time constant of C and R_1 is long compared to the input pulse duration; hence, C charges only slightly more for the duration of the pulse from t_2 to t_3, and current through the resistor is low. If, for instance, the capacitor charge rises to 3 V at the end of the pulse duration, the output pulse amplitude declines by 1.5 V to a value of 4.5 V at t_3, as shown in Fig. 3-23b.

At time t_3 when the input pulse amplitude drops, the voltage across the resistor becomes $-3 + 2.5 = -0.5$ V. Hence, the output voltage becomes -0.5 V at t_3, producing a short negative spike in the output waveform, as shown in Fig. 3-23b. This negative voltage across the diodes causes conduction, and the low impedance of the conducting diode shunts R_1, changing the long time constant to a short one for a brief interval. Consequently, a rapid discharge occurs for the -0.5-V charge of the capacitor. The capacitor discharge through the diode for -0.5 V leaves the original 2.5-V charge on the capacitor. Since this charge is a static condition for the time interval between pulses, there is no current flow through R_1 and, hence, no potential drop across it. Thus, the interval between the pulses to the right of the capacitor is clamped to zero level (Fig. 3-23b).

On occasion it may be necessary to clamp the output voltage level at some value above or below 0. When this is the case, the circuit is modified by the inclusion of a fixed potential (battery or other power source), as shown in Fig. 3-23c. A bypass capacitor C_2 across the power source places the bottom of the resistor and diode at *signal* ground by virtue of the bypassing effect of the low reactance.

Assume that the pulse train is to be clamped at 1 V above the 0 line. In such an instance the voltage source is positive toward the diode anode (Fig. 3-23c). At time t_1 the charge across the coupling capacitor C_1 would normally be a fixed 2.5 V, but the 1-V bias will oppose the charge voltage and, consequently, the capacitor C_1 charges to only -1.5 V.

At time t_2 the circuit function is similar to that described for the circuit shown in Fig. 3-23b. The rapid rise of the pulse amplitude is too quick for the long time constant of the RC network to permit any charge to build up across the capacitor during the leading-edge time interval. Thus, full pulse current flows through resistor R_1 causing 5 V to develop across it. Electron flow is up through the resistor, and the polarity of the voltage drop across R_1 coincides with the bias polarity, producing a voltage rise at the output to 6 V (5 V + 1 V). During the rapid rise time, C_1 and C_2 have little charge built up across them, but permit full current flow through R_1.

Assume RC is such that during the flat-top interval of the pulse an additional 1-V charge develops across capacitor C_1. This additional charge raises the charge across C_1 to -2.5 V again, since it had a -1.5-V charge on it originally. Now the output pulse amplitude at the base of Q_2 drops to 5 V.

At time t_3 when the input pulse drops in amplitude, there is an excessive negative charge of 1 V across C_1, permitting the diode to conduct. During conduction the excessive 1 V is discharged and the C_1 charge reverts to -1.5 V again (2.5-V source, minus the 1-V bias). The discharge produces the negative spike to the 0 line (Fig. 5-23c). If the excessive charge on the capacitor had been only -0.5 V, the spike would not have reached the 0 reference line. In either event, however, the waveform is clamped to the 1-V level, as shown.

questions and problems

3-1. Explain briefly what is meant by the *piezoelectric* effect.

3-2. For a quartz crystal, which axes are known as the mechanical type?

3-3. Determine the thickness required for the following crystal cuts for resonance at 1000 kHz: *X, Y, AT*, and *BT*.

3-4. A quartz crystal is resonant at 500 kHz and its equivalent inductance (L) is 3.2 H. If the equivalent resistance is 8500 Ω, what is the Q?

3-5. Briefly explain why the output resonant circuit of a crystal oscillator must be tuned slightly higher in frequency than the resonant frequency of the crystal.

3-6. If the inductance for a resonant circuit is 4.1 μH and the capacitance is 0.0002474 μF, what is the resonant frequency and wavelength?

3-7. If the product of LC is 0.1340, what is the resonant frequency in kilohertz and the wavelength in meters?

3-8. If the frequency of a resonant circuit is 5 MHz and the capacitance value is 0.00025 μF, what must be the value of the inductance?

3-9. If the inductance has a value of 2.5 μH and the resonant frequency is 60 MHz, what must be the value of the capacitance?

3-10. What are some of the advantages of the MOSFET over tubes and ordinary transistors?

3-11. Briefly describe the circuit differences between a Hartley oscillator and a Colpitts.

3-12. Why is the electron-coupled oscillator superior to the ordinary tuned RF oscillator?

3-13. What are the two basic types of relaxation oscillators and what are the essential circuit differences between them?

3-14. Briefly explain how a relaxation oscillator can be synchronized by an external signal.

3-15. How does a discharge circuit form sawtooth signals? Explain briefly.

3-16. If $C = 0.003 \mu F$ and $R = 120,000 \Omega$, what is the time constant in microseconds?

3-17. If the capacitance is $0.005 \mu F$ and the time constant is $300 \mu s$, what is the resistance value in ohms?

3-18. In a discharge circuit the capacitor must charge to 14 per cent of applied voltage in $345 \mu s$. If the resistance is $50,000 \Omega$, what must be the value of the capacitance?

3-19. A capacitor must charge to 10 per cent of its full value in $100 \mu s$. If the capactior has a value of $0.005 \mu F$, what must be the value of the series resistance?

3-20. A 200,000-Ω resistor is in series with a 0.00025-μF capacitor and a 30-V source. After $50 \mu s$ what is the instantaneous voltage across the capacitor? [Use Eq. (3-11).]

3-21. Each pulse in a series has a duration of $5 \mu s$ and the pulses occur 2650 times per second. What is the duty cycle?

3-22. If the peak power for the pulses of Problem 3-21 is 27 W, what is the average power of the pulse train?

3-23. If a pulse in a series has a duration of $5 \mu s$, what is the base frequency and what is its significance?

3-24. What are the essential differences between an RC integration circuit and an RC differentiating circuit?

3-25. How does the LR integration circuit differ from the RC and which is more advantageous? Explain.

3-26. Explain the useful features of series and shunt clippers.

3-27. Explain what is meant by the *dc level* of a pulse train and what could cause its loss.

3-28. Briefly explain how a clamper restores the dc level to a train of amplified pulses.

4

RF and IF amplification

4-1. introduction

After the carrier signal has been generated in transmitting systems, it is necessary to provide RF amplifier stages to bring the signal power level to that desired for application to the antenna systems, as shown in Figs. 4-1a and 2-1. Such RF stages are usually resonant-circuit types, which receive a high degree of driving signal power from the previous stage (called excitation). The amplifiers may be tuned to the frequency of the incoming signal (from an oscillator or a previous RF amplifier), or tuning may be to some multiple of the frequency of the input signal (see Fig. 3-1b).

In receivers RF amplifiers are also necessary, as shown in Fig. 4-1b, so that weak signals obtained from the tuner system (see also Fig. 3-8) can be brought to proper amplitudes for application to the demodulating systems. Such amplifiers are usually termed *intermediate-frequency* amplifiers (IF), since they handle a frequency resulting from the heterodyning of the carrier signal and that of a local oscillator, as more fully described in Chapter 6.

Often an RF amplifier is also employed before the mixer-oscillator section of a receiver for greater selectivity and signal sensitivity (see Fig. 3-8). All these types are covered in this chapter.

4-2. classes of amplifiers

Radio-frequency (RF) amplifiers (as well as the audio types discussed in Chapter 5) may be operated as Class A, AB, or B, as required to meet specific operational characteristics. A fourth category of amplification, Class C, is

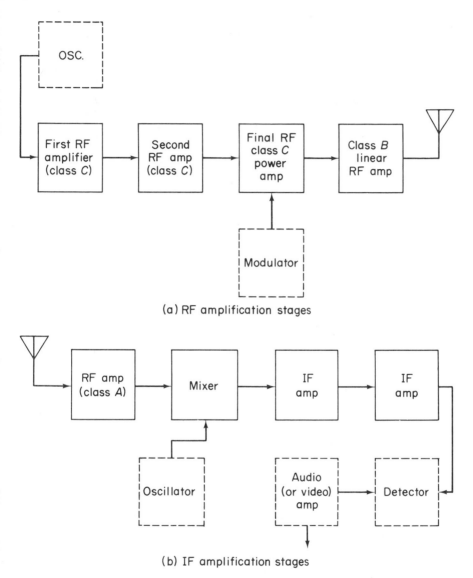

(a) RF amplification stages

(b) IF amplification stages

figure 4-1 RF and IF amplifier applications

used only for RF power amplification of unmodulated carrier signals in transmitters, as discussed more fully later.

Graphs indicating bias and signal requirements for the various classes of amplifiers as shown in Fig. 4-2. While these are the conventional methods for illustrating the classification of amplifiers, signal current changes and polarities are applicable only to a tube-type amplifier with a signal-grounded

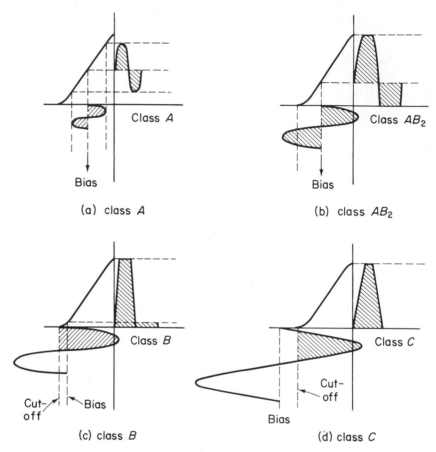

(a) class A

(b) class AB₂

(c) class B

(d) class C

figure 4-2 types of amplification

cathode of the type shown in Fig. 4-3a. Such an amplifier can have resonant circuits at the input and output sections instead of the resistors shown. Negative bias may be applied to the bottom of R_1 instead of having the latter at ground. For the circuit shown, bias is developed by R_2, since current flow through this resistor places the cathode at a positive potential greater than the grid, thus making the grid minus with respect to the cathode. Capacitor C_1 effectively places the cathode at *signal ground.*

For Class A operation, the bias is set about midway up the linear portion of the tube's characteristic curve, so some idling current flows without a signal. The input signal amplitude is kept within bounds so that it does not extend into the curved portions of the characteristic curve, as shown in Fig. 4-2a.

For Fig. 4-3a a positive alternation of the input signal decreases grid bias and increases plate-current flow. There is a larger voltage drop across resistor

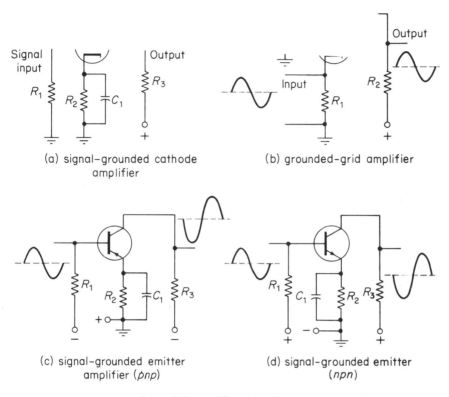

(a) signal–grounded cathode amplifier

(b) grounded-grid amplifier

(c) signal-grounded emitter amplifier (*pnp*)

(d) signal-grounded emitter (*npn*)

figure 4-3 amplifier phase factors

R_3 and, consequently, the output signal voltage drops, causing a phase reversal between input and output voltages. The input grid *voltage* signal and the plate-*current* changes are graphed at the upper left of Fig. 4-2 and represent Class A operation.

For the amplifier shown in Fig. 4-3b, the grid is grounded and signal input is applied across the cathode resistor. Here a positive alternation of the input signal voltage *increases* bias, since it will make the cathode more positive than the grid. Now plate current drops and there is a decrease in the voltage drop across R_2, producing a rise in output signal voltage. Thus, even though this grounded-grid amplifier circuit operates on the linear portions of the characteristic curve, the phase and current relationships are not as graphed in Fig. 4-2.

For the amplifier shown in Fig. 4-3c, a negative voltage is applied to the base, thus furnishing the necessary forward bias to permit conduction. Bias on the collector, however, is negative with respect to the emitter (reverse bias) in contrast to vacuum-tube operation. Hence, when the input signal has a positive alternation, forward bias is decreased and collector current declines, with less drop across the output resistor R_3. There is a voltage increase in the negative direction toward the supply potential. Thus, the output current change differs from that of the graphs.

For the circuit shown in Fig. 4-3d, the collector potential is positive, thus furnishing reverse bias to the collector emitter of the *npn* transistor. While a positive collector potential is comparable to the positive voltage applied to a tube's anode, the input potentials differ here, since the forward bias at the base emitter must be positive for the base. Thus, a positive alternation of the input signal now increases the forward bias, thus increasing collector current flow. Voltage across R_3 rises and collector voltage drops. There is again an out-of-phase condition between the input and output signal voltages, though current changes again differ from those graphed.

While transistor circuitry differs, the basic characteristics indicated in Fig. 4-3 still apply regarding amplifier classification. When the input signal is such that saturation is reached for one of its alternations and cutoff prevails for the other alternation, as shown in Fig. 4-2, the operation is Class AB_2, with the subscript indicating that, for a tube, grid current flows because the grid becomes positive during the peaks of the positive alternations of the input signal. If the input signal had only sufficient amplitude to reach saturation and cutoff without extending into these regions, the classification would be AB_1.

Class B, shown in Fig. 4-2, is characterized by a bias so set that only the positive alternations of the input signal cause current flow in the output circuit. Sometimes bias is set slightly below the cutoff point, at the place on the graph called the *projected* cutoff. This is determined by projecting the linear portion of the characteristic curve to the horizontal axis and setting the bias point there so that operation will be more linear.

Class AB and B amplifiers provide more output than Class A because of increased efficiency, though harmonic distortion possibilities are much greater. For RF amplification, where resonant circuits are present, operation may be with a single tube or transistor (*single-ended*) or push–pull, described later, which increases power additionally and reduces distortion. For audio operation as described in Chapter 5, single-ended operation may be employed for Class A or AB, though Class B requires push–pull operation to avoid severe distortion.

Class C, shown in Fig. 4-2, provides the greatest efficiency and power output, though it is suitable only for RF amplification, since only a portion of the input signal is used and audio information would be lost. As shown, bias is set beyond the cutoff point and a high-amplitude driving signal is required for proper operation, as more fully discussed later.

The efficiency of an amplifier is determined by the exact bias employed, the amount of drive furnished by the input signal, and other design factors. Generally, however, Class A operation has an efficiency between 20 to 25 per cent at full output. Class AB_2 operation extends efficiency into the 35 to 50 per cent region, while efficiency for Class B in vacuum-tube operation may range from 60 to 70 per cent. Class C amplifier efficiency is over 90 per cent in well-designed circuits for RF amplification.

Bias factors and the amount of driving signal at the input also apply to transistor circuitry. Generally, Class A operation is achieved by utilizing forward bias between base and emitter, and reverse bias between collector and emitter. The input signal amplitude is kept small enough to operate linearly with a minimum of distortion.

For some transistors, cutoff is obtained when no forward bias is applied to the base-emitter circuitry, or by making the base slightly positive for the *pnp* types (or slightly negative for the *npn* types) to provide a small degree of reverse bias. Thus, Class B operation can be obtained. For Class C, reverse bias is applied to both the input and output sections of the transistor, thus providing for a bias beyond the cutoff point. Other solid-state amplifiers can also be operated in the class desired. With the FET, for instance, the gate bias is adjusted to meet the classification required.

4-3. resonant-circuit factors

The use of series- or parallel-resonant circuits in RF amplifiers permits the selection and amplification of desired signals and the rejection of undesired signals. The ability of an amplifier to discriminate against undesired signals is termed the *selectivity* of the system. An indication of an amplifier's selectivity can be determined by plotting the current flow in the output circuit against frequencies extending from below to above the resonant frequency, as shown in Fig. 4-4.

As a series circuit is tuned to resonance, current rises, since the inductive and capacitive reactances cancel, leaving only resistance. For a parallel-resonant circuit, however, current drops as circuit impedance rises. Even though we designate a circuit as having a specific resonant frequency, other signals with frequencies slightly above and below the resonant frequency will also be amplified virtually as much as the signal of resonant frequency. Thus, resonant circuits usually are designed to pass a band of signals with frequencies around the resonant one.

The bandpass characteristics of a circuit are established by the f_1 and f_2 points shown in the upper curve of Fig. 4-4. These are termed the *half-power* point frequencies and are qual to 0.707 times the peak amplitude. Thus, they are the frequencies at which the amplitudes of both the high- and low-frequency slopes of the resonant curve are 0.707 of the peak amplitude.

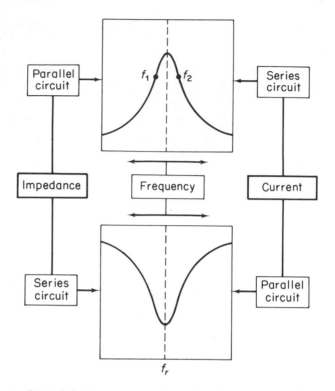

figure 4-4 resonant curves for series and parallel circuits

Thus, the f_1 and f_2 points define the selectivity of the circuit and indicate the bandpass:

$$\text{bandwidth} = f_2 - f_1 \qquad (4\text{-}1)$$

The particular bandpass used depends on the requirements of the system in question in terms of the amount of signal frequencies to be transmitted and received for conveying the information. Radio receivers, for instance, have a bandpass of approximately 10 kHz, while for FM each station is allocated a 200-kHz bandwidth. In TV broadcasting each station is allocated a 6-MHz bandwidth, which also includes a 50-kHz section for the sound portion of the telecast.

The shape and bandpass characteristics of a resonant-circuit curve can be altered by changing circuit resistance or the L-to-C ratio. With a higher L/C ratio, selectivity increases (the curve narrows) even though the resonant frequency remains the same. (A specific product of LC produces a given frequency, even though the L and C values are changed.) For a series circuit an increase in series resistance lowers selectivity. For a parallel circuit, an increase in shunt resistance raises selectivity.

In a series-resonant circuit the inductor is most likely to have some resistive component, even though heavy gauge wire is used. With air-dielectric tuning capacitors, however, leakage resistance is virtually zero. Thus, inductor resistance must be lumped with whatever other series resistance is present in the resonant circuit to ascertain circuit selectivity. The figure of merit that refers to circuit selectivity is designated by Q and has already been mentioned in Section 3-2. Equation (3-3) also applies here:

$$Q = \frac{6.28fL}{R} = \frac{X_L}{R} \tag{4-2}$$

Thus, selectivity can be increased in a series-resonant circuit by increasing the inductive reactance (and at the same time altering C to maintain the same LC product), by decreasing the resistance, or by both increasing L and decreasing R.

Since the bandpass is related to selectivity, Q is also related to the $f_2 - f_1$ points given in Eq. (4-1). Hence, the following mathematical relationships prevail for the series-resonant circuit:

$$\Delta f = f_2 - f_1 = \frac{f_r}{Q} = \frac{R}{2\pi L} \tag{4-3}$$

Dividing both sides by f_r (resonant frequency) yields

$$\frac{\Delta f}{f_r} = \frac{f_2 - f_1}{f_r} = \frac{1}{Q_r} \tag{4-4}$$

$$f_2 = f_r + \frac{R}{4\pi L} \tag{4-5}$$

$$f_1 = f_r - \frac{R}{4\pi L} \tag{4-6}$$

$$Q = \frac{f_r}{f_2 - f_1} \tag{4-7}$$

$$f_r = Q(f_2 - f_1) \tag{4-8}$$

Knowing the values of L, C, and R, we can solve for Q by

$$Q = \frac{1}{R}\sqrt{\frac{L}{C}} \tag{4-9}$$

As evident from an inspection of the foregoing equations, resonant circuits can be designed for the bandpass required to handle the carrier and sideband signals comprising the composite signal transmitted. Successive amplifier stages can also be tuned to different resonant frequencies so that a desired overall bandpass is achieved, as shown more fully later.

As an illustration of the use of various equations for selectivity, assume that a series-resonant circuit has a resistance of 10 Ω, an inductance of 50 μH and a capacitance of 0.0002 μF. What is Q, bandwidth, f_r, f_2, and f_1?

In solving this problem we use Eq. (4-9) initially:

$$Q = \frac{1}{10}\sqrt{\frac{50}{0.0002}} = \frac{1}{10}\sqrt{25,000} = 0.1 \times 500 = 50$$

Next we solve for bandwidth using Eq. (4-3):

$$\text{bandwidth} = \frac{R}{6.28 \cdot L} = \frac{10}{6.28 \cdot 50 \cdot 10^{-6}} = 31.847 \text{ kHz}$$

Now we can solve for the resonant frequency by Eq. (4-8):

$$f_r = Q(f_2 - f_1) = 50 \cdot 31.847 \cdot 10^3 = 1592.35 \text{ kHz}$$

Equations (4-5) and (4-6) are now used to find the f_1 and f_2 values:

$$1592.35 \cdot 10^3 + \frac{10}{12.56 \cdot 50 \cdot 10^{-6}} = 1592.35 \cdot 10^3 + 15.92 \cdot 10^3 = 1608.27 \text{ kHz}$$

Note that the $R/12.56 \cdot L$ value is one half the bandwidth value obtained earlier. Thus, when using Eq. (4-6) for obtaining the f_1 value, the 15.92 kHz is subtracted from the resonant frequency to obtain 1576.43 kHz.

Having solved for the various frequency values, we can prove the Q value by using Eq. (4-7):

$$Q = \frac{f_r}{f_2 - f_1}$$

$$= \frac{1592.35 \cdot 10^3}{(1608.27 \cdot 10^3) - (1576.43 \cdot 10^3)}$$

$$= \frac{1592.35 \cdot 10^3}{31.84 \cdot 10^3} = 50$$

In circuit design practices the particular equations are selected that are most convenient for solving the problem involved. Different approaches are possible, some providing answers with less mathematical manipulations than others. In the previous example, for instance, we could have solved for f_r initially, using Eq. (3-4), followed by use of Eq. (4-2) for obtaining Q:

$$f_r = \frac{1}{6.28\sqrt{LC}} = \frac{1}{6.28\sqrt{50 \cdot 0.0002}} = \frac{1}{6.28 \cdot 0.1} = 1592.35 \text{ kHz}$$

$$X_L = 6.28fL = 6.28 \times 1592.35 \times 10^3 \times 50 \times 10^{-6} = 500 \, \Omega$$

$$Q = \frac{X_L}{R} = \frac{500}{10} = 50$$

As mentioned, decreasing R or increasing L will increase selectivity and provide for a more narrow bandpass. Thus, if we double the inductance value previously given so that it becomes 100 μH and halve C to keep the same LC product and frequency, we obtain

$$X_L = 6.23 \times 1592.35 \times 10^3 \times 100 \times 10^{-6} = 1000 \, \Omega$$

$$Q = \frac{X_L}{R} = \frac{1000}{10} = 100$$

Now that the Q has been raised to 100, we find the bandpass halved:

$$\frac{R}{6.28 \times 100 \times 10^{-6}} = 15.92 \text{ kHz}$$

For parallel circuits at resonance, the bandwidth equation (4-1) also applies. The Q would be determined by the value of the shunting resistance. For practical purposes, the shunting resistance would be the load resistance applied to the resonant circuit for obtaining the RF power developed. If R is considered as the load resistance applied in parallel to the resonant circuit, the equation for Q becomes

$$Q = \frac{R}{X_L} \qquad (4\text{-}10)$$

For an unloaded parallel resonant circuit with negligible resistance in the inductor, the Q becomes

$$Q = \frac{Z}{X_L} \qquad (4\text{-}11)$$

The second portion of Eq. (4-3) also applies, where $f_r/Q = f_2 - f_1$.

In communication design practices, bandpass characteristics can be altered by RF filters composed of series and/or shunt inductors and capacitors, or sections of RF transmission lines can be used for such purposes. These circuits are discussed in detail in Chapter 10.

4-4. grounded-base amplifier

A typical grounded-base amplifier (also termed *common base*) is shown in Fig. 4-5. An input transformer is used (L_1 and L_2) with C_1 forming a resonant circuit between the emitter and base of the *pnp* transistor. An *npn* could, of

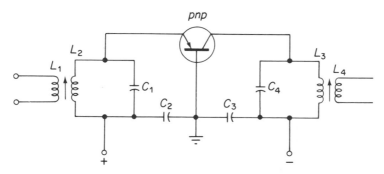

figure 4-5 grounded-base RF amplifier

course, also be used with polarity reversal of supply potentials. The arrow between the L_1 and L_2 windings denotes an adjustable core (tuning slug). Variable-capacitor tuning could also be employed, as shown in subsequent diagrams. A similar resonant circuit and tuning arrangement are present at the output between the base and collector, comprising inductance L_3 and L_4, as shown.

The forward bias positive potential is applied to the emitter of the *pnp* transistor through L_2, with C_2 placing the bottom of the resonant circuit at *signal ground* by virtue of the low reactance of C_2 for the resonant frequency signals. The reverse bias (negative to the collector) is also applied through a resonant-circuit inductor, L_3 in this case. Again a capacitor (C_3) is used to place the bottom of the resonant circuit at signal ground.

The grounded-base circuit is characterized by an in-phase condition between the input signal and the amplified version appearing at the output. Compared to the common-emitter amplifier discussed next, the common base is less affected by low-frequency oscillations, since it has less gain at frequencies below f_r than the common emitter. Also, the signal frequencies above the resonant frequency extend to higher regions before they tend to fall off, thus facilitating wide-band operation when this is required. With a common-base RF amplifier system the output circuit is isolated from the input by the grounded base; hence, there rarely is need for neutralization. This is not the case with the common-emitter circuitry.

4-5.   common-emitter-type amplifier

An RF common-emitter (grounded-emitter) amplifier is shown in Fig. 4-6. This could be an IF amplifier in a superheterodyne receiver, in which case L_1 receives the input signals from the mixer output. The same circuit could also represent the RF stage of a receiver (feeding the mixer stage). In the latter instance, L_1 and L_2 would be associated with the antenna network. As an IF amplifier, capacitors C_1 and C_5 would be of the trimmer type only, used to adjust for resonance within a small range around f_r. In RF amplification the tuning capacitor is a standard variable and ganged to that of the mixer and oscillator stages (see Fig. 3-8).

The terminal designated AVC is for automatic-volume-control purposes. A voltage is obtained from the demodulator system and applied to R_1 for purposes of varying the forward bias of the transistor (or grid bias of a vacuum-tube amplifier) for regulating the degree of gain in proportion to the amount of signal input. This prevents overloading from local stations and automatically maintains the volume level in radios at that set by the user by adjustment of the volume control. Circuitry and additional discussion of

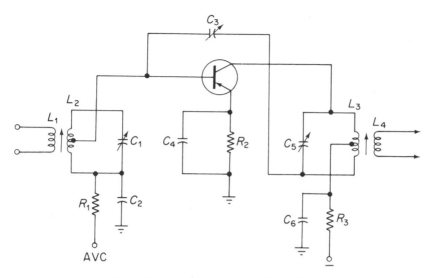

figure 4-6 grounded-emitter RF amplifier

AVC are given in Chapter 6. In television receivers similar functions are termed *automatic gain control* (AGC), as discussed in Chapter 8.

For Fig. 4-6, capacitor C_2 places the bottom of the resonant circuit at signal ground, as was the case for the common-base circuit. Capacitor C_4 and resistor R_2 are for stabilizing purposes. If temperature changes cause an increase in current through the collector-base circuitry, the base side of resistor R_2 becomes more negative, thus decreasing the potential difference between the collector and emitter and lowering current flow in a compensatory manner.

In the common-emitter circuit there is a phase reversal between the input and output signals, plus the tendency to oscillate because of common interelement capacitances (or interelectrode capacitances in tubes). To eliminate the oscillations at the resonant frequency of the amplifier, a neutralizing capacitor C_3 is used from the bottom of the output resonant circuit to the base, as shown in Fig. 4-6.

Note that L_3 is tapped by C_6 and R_3, with the capacitor placing the tap at signal ground. Thus, the signal at the bottom of the resonant circuit is out of phase with that at the collector side, and a portion of this is fed back by C_3 to cancel the coupling effects of the interelement capacitances causing oscillations. The neutralizing capacitor is tuned so its reactance provides the proper degree of feedback voltage.

For solid-state circuitry the word *unilateralization* has become a common designation for neutralization particularly where both the R and X factors of internal feedback are cancelled. Thus, a unilateral circuit is actually a one-way path for the signal. When an input signal is applied to the input, an amplified replica appears at the output. If, however, a signal is applied to the output circuitry, no path in reverse to the input is present and the signal does not reach the input side.

4-6. FET RF circuit

An RF amplifier using a *field-effect transistor* (FET) is shown in Fig. 4-7. Typical values for operation from the 100- to 200-MHz region are given for reference purposes. For comparison purposes the applied voltages shunt the resonant circuits; hence, power-supply current does not flow through them. This design is termed *shunt feed* in contrast to the *series feed* shown for the previous amplifiers in this chapter. For series feed the supply current flows through the resonant-circuit inductor (or load resistor, if one is used).

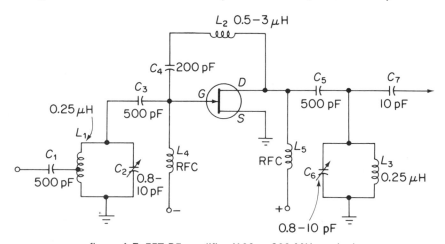

figure 4-7 FET RF amplifier (100 to 200 MHz region)

As shown in Fig. 4-7, an RF choke (L_4) is in series with the supply potential and the gate of the FET. This RFC provides a high reactance for signals and prevents their leakage to the supply system. Capacitor C_3 blocks the direct current and prevents the supply voltage from being grounded through the low resistance of L_1. A similar arrangement prevails at the drain side, where C_5 blocks the supply voltage applied to the bottom of the RFC L_5. Capacitors C_1 and C_7 also prevent shorting of direct current at the input and output circuits. Capacitor C_1 taps L_1 where the proper impedance prevails for matching purposes.

Unilateralization is provided by inductor L_2 and capacitor C_4. The latter also prevents direct coupling of direct current between the gate and drain of the FET. The degree of feedback is regulated by adjustment of the values of the series inductor and capacitor. As these elements are tuned closer to the resonant frequency, the impedance of the series network drops and more RF energy is fed back for neutralization purposes. The impedance rises for off-resonance conditions (see Fig. 4-4) and less signal voltage is transferred.

4-7. coupling effects

When successive circuits are coupled to each other by the transformer arrangements shown in Figs. 4-5 and 4-6, the bandpass characteristics of the individual resonant circuits are altered because of the influence of the coupled circuits. The degree of coupling (how close the primary and secondary are placed with respect to the interaction of their individual fields) not only influences the bandpass, but also the gain, as shown in Fig. 4-8.

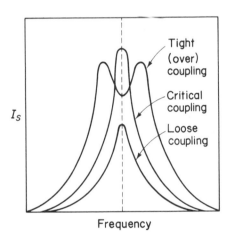

figure 4-8 coupling effects on response curve

With loose coupling, the secondary current has a low amplitude and is sharply peaked. As the primary and secondary coils are brought into closer proximity, more of the load resistance is reflected back into the primary system. The current in the secondary peaks at a higher level and the bandpass broadens, as shown in Fig. 4-8. When the coupling has been increased to the point where the resistance reflected into the primary is equal to the primary resistance (impedances match), the degree of coupling is known as *critical*. Now the secondary current reaches its maximum value.

If the coupling is increased additionally, it becomes known as *tight coupling* or *overcoupling*. The response curve undergoes a dip around the resonant frequency, as shown in Fig. 4-8. Thus, the gain for the resonant frequency signal has been decreased and new resonant peaks occur each side of center. Such coupling is utilized on occasion for obtaining a broad band-pass, as discussed more fully later.

When inductors are placed into magnetic-field proximity to form a transformer, the overall inductance is the same as that of inductances placed in series aiding (that is, the individual coils are wound in the same direction so that their magnetic fields do not oppose each other). (RF energy is transferred via the inductances only, even though a resonant-circuit shunt capacitor is present.)

The additional inductance resulting from the coupling of two coils is termed the *mutual inductance* (symbol M). When an ac signal of 1 A flowing in the primary induces 1 V across the secondary, the two inductances have a mutual inductance of 1 H. When all the magnetic lines of force of the primary cut the secondary, the formula for mutual inductance is

$$M = \sqrt{L_1 L_2} \tag{4-12}$$

where M = mutual inductance in henrys
$\quad L_1$ = primary inductance in henrys
$\quad L_2$ = secondary inductance in henrys

For the resonant circuits of RF amplifiers, the coupling is not such that all the magnetic lines of force of the primary winding cut the secondary. Hence, the mutual-inductance formula is modified to take into consideration the degree of coupling, known as the *coefficient of coupling* and indicated by the symbol k. The coefficient of coupling represents the *percentage* of coupling. Thus, a k of 0.5 means only half the lines of force of the primary intercept the secondary winding; hence, the coefficient of coupling is actually 50 per cent. The k is added to Eq. (4-12) to form

$$M = k\sqrt{L_1 L_2} \tag{4-13}$$

As an example, assume that a transformer primary has an inductance of 8 μH and a secondary inductance of 2 μH with a coefficient of coupling of 0.6. What is the mutual inductance?

$$M = 0.6\sqrt{2 \times 8} = 0.6\sqrt{16} = 2.4 \ \mu\text{H}$$

In a series circuit the total inductance L_T is

$$L_T = L_1 + L_2 + 2M \tag{4-14}$$

For the example given, the total inductance is $2 + 8 + (2 \times 2.4) = 14.8$ H. This is of academic interest only, however, since the individual inductances plus their shunting capacitors make up the individual resonant circuits.

If the mutual inductance plus individual inductances are known, k can be found by

$$k = \frac{M}{\sqrt{L_1 L_2}} \tag{4-15}$$

As with Q factors discussed earlier, bandpass is established by the Q designed into the system. At critical coupling the coefficient (k_c) is related to the individual circuit Q of the primary and secondary by the equation

$$k_c = \frac{1}{\sqrt{Q_p Q_s}} \qquad (4\text{-}16)$$

where k_c = coefficient of critical coupling
Q_p = Q of primary winding
Q_s = Q of secondary winding

For wide-band operation in television and frequency-modulation systems, circuit selectivity is altered to procure the necessary bandpass. While overcoupling increases bandpass, the dip at the resonant frequency, plus the two adjacent peaks, makes for an uneven response to the various signals. A response curve with a substantially flat top can be obtained by using two or more amplifier states with different degrees of coupling. Thus, the first amplifier could have tight coupling and the second amplifier critical coupling, as shown in Fig. 4-9a. The result is the combination of the two curves to pro-

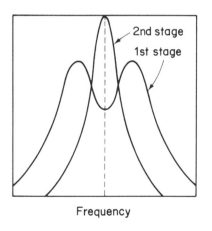

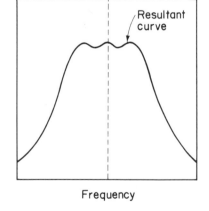

Frequency   Frequency

(a) coupling of each stage   (b) resultant response curve

figure 4-9 effect of double-tuned circuits on curves

duce the overall response curve shown in Fig. 4-9b. Now the slight dips in the flat top are negligible if the total curve amplitude is fairly high.

Similar results of wide-band operation with a flat-top response curve can also be obtained by the system known as *stagger tuning*. This is accomplished by tuning successive stages to slightly different frequencies around the resonant point. Again, the combined amplifier responses result in a single wide-band curve of the type shown in Fig. 4-9b.

On occasion (such as in AM and FM receiver combinations) double-transformer design is used, such as shown in Fig. 4-10. Here a common

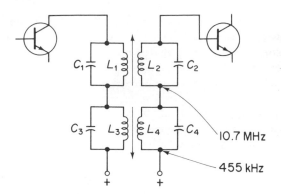

figure 4-10 dual IF coupling

amplifier is used for both the IF for FM (10.7 MHz) and that for AM (455 kHz). When the receiver is tuned to FM, the signal applied to the IF stages will be 10.7 MHz, which will cause the upper circuit (now at resonance for the 10.7 MHz) to have a high impedance, as is characteristic for a parallel circuit. Thus, the signal will develop a high amplitude across this resonant circuit and normal amplification occurs. The lower circuit, however, not being at resonance, will have a low impedance for this frequency and will act as a low resistance path for completion of the signal circuitry to the emitter section.

When the receiver is tuned to AM (assuming an IF of 455 kHz) the lower circuit is at resonance and offers a high impedance. Hence, a large-amplitude signal voltage develops across it. The upper circuit, however, is not at resonance and, hence, has a low impedance. It thus offers a low resistance closed circuit for the signal energy.

Capacitors C_1, C_2, C_3, and C_4 are usually fixed in value, with tuning adjustments made by resetting the placement of the metallic cores of the transformers. The coils are mounted in a metal can for shielding, with one core adjustment at the top and the other at the bottom, identified by the position shown by the arrow heads (Fig. 4-10). Since IF's are fixed-frequency sections, tuning is only necessary on rare occasions. Actual station tuning is made in the tuner section of the receiver with either variable (ganged) capacitors or variable-core inductors linked to the tuning knob.

4-8.   class B linear amplifier

A Class B linear RF amplifier is shown in Fig. 4-11. Such an amplifier is useful in communication systems for amplification of a *modulated* RF carrier signal. Once a carrier has been amplitude-modulated in a Class C amplifier stage, successive Class C circuits are unsuitable for additional amplification since they are unable to handle a modulated signal properly. Class C stages will not reproduce all the modulation components of the composite carrier since plate current flows for only a portion of the positive-going alternations of the input signal, as shown in Fig. 4-2. In the Class B amplifier, however, all the positive alternations of the input signal affect plate current flow, because the bias is at approximately the cutoff point and not beyond it, as is the case with Class C. Since resonant circuits are present, the missing alternations of the input signal are reintroduced because of the flywheel effect present.

For the amplifier in Fig. 4-11, a pentode tube is utilized, though a triode

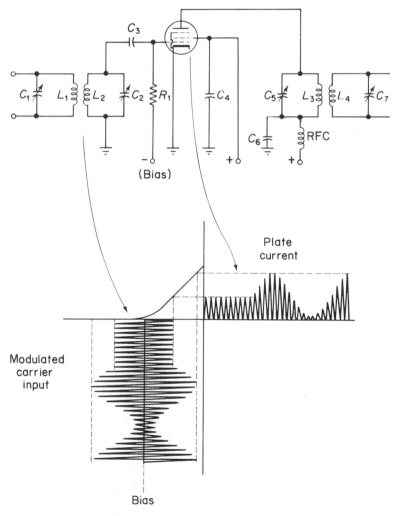

figure 4-11 class B (linear) RF amplifier

could also be employed if appropriate neutralization is introduced into the circuitry, as discussed for the grounded-emitter RF amplifier. With a pentode tube specifically designed for the frequency span to be utilized (and not operating at too high a frequency), neutralization may not be necessary. In a pentode the screen grid can be placed at signal ground with a bypass capacitor (C_4), thus providing good isolation between the input and output circuits in terms of interelectrode capacitive coupling. In addition, the suppressor grid (between the anode and screen grid) is also at ground potential because it is tied in to the cathode. The minimization of capacitances reduces the tendency toward self-oscillation. Interelectrode capacitances in well-designed pentodes may only be a few picofarads, but if operated at a sufficiently high signal frequency, some reactances will be sufficient to cause oscillations, particularly since the gain and power sensitivity of pentodes is so high that only a very small feedback signal is needed. Thus, neutralization may be necessary in some applications.

For Fig. 4-11 the resonant circuit composed of C_1 and L_1 represents the output of the previous-stage Class C amplifier. The modulated carrier appearing across this resonant circuit is coupled to the Class B stage by the transformer composed of L_1 and L_2.

For Class B operation, a fixed bias is essential; hence, a negative potential is applied to the bottom of the grid resistor R_1, as shown. Bias from a cathode resistor would be unsuitable, since virtually no anode current flows during the absence of a grid driving signal; hence, no bias would be developed during this time. In low-signal amplification in vacuum-tube usage (Class A or AB_1), some idling current flows in the anode circuit with or without an input signal, and bias from a cathode resistor is thus feasible in such instances. In transmitting circuitry, however, the cathode resistor also consumes considerable power and also reduces the available plate potential. The bias for the amplifier of Fig. 4-11 is blocked with C_3 to prevent shorting it to ground through L_2.

As shown in the waveform below the drawing in Fig. 4-11, the excitation applied to the grid of the Class B linear amplifier is such that the positive alternations of the input carrier signal (when not amplitude modulated) have a peak that extends to the midpoint between bias cutoff and the zero-bias point of the tube. By setting the excitation amplitude to this level, the carrier can increase and decrease during the modulation process to equal proportions above and below the unmodulated level without exceeding the limits of cutoff and the region just before tube saturation. In practical applications the maximum positive amplitude of the unmodulated carrier is set at approximately the center of the linear portion of the characteristic curve.

Since the negative alternations of the grid input signal extend into the tube's cutoff region, plate current only flows in pulses, as shown. As mentioned, however, the flywheel effect of the resonant circuit in the plate side (C_3

and L_3) reintroduces the missing alternations to reform the amplified modulated carrier at the output of the Class B. Since the positive alternations of the input signal vary between cutoff and near the saturation region, instead of to saturation for each alternation, as in the case of the Class B discussed in Section 4-2, efficiency is approximately 30 per cent instead of the approximately 65 per cent mentioned for the Class-B amplifier handling unmodulated RF only.

As with other amplifiers discussed previously, the radio-frequency choke (RFC) in series with the power supply and the resonant circuit in the anode isolates the amplified RF energy from the power supply and thus minimizes losses. Capacitor C_6, by providing a low-reactance path to ground (and thus back to the tube's cathode), also helps confine the RF energy to the cathode–anode section of the output circuitry.

The transformer-coupled resonant output circuit composed of L_4 and C_7 couple the amplified modulated carrier to another Class B linear stage if added amplification is necessary, or to the antenna system for propagation purposes.

4-9. class C amplifiers

As mentioned in Section 4-2, the Class C amplifier provides for more RF output and efficiency than other type of amplifiers. Hence, it is extensively used in transmitting systems for increasing RF carrier power to the desired level. Both the transistor and tube versions are shown in Fig. 4-12. The solid-state devices are ideal for portable use when low or intermediate power levels are utilized. For high-power applications in broadcast stations, tubes capable of handling thousands of watts of carrier power are utilized in the final stages of carrier amplification.

For the circuit shown in Fig. 4-12a, the unmodulated RF input is applied to the input circuit of the amplifier and appears across the parallel-resonant circuit consisting of L_2 and C_2. Inductance L_2 is tapped to decrease the normally high resonant-circuit impedance to that which matches the emitter-base input of the transistor. The necessary forward bias is supplied by a separate bias pack to the bottom of the input resonant circuit, as shown. Bias for the Class C is set at two to three times cutoff (see Fig. 4-2). Resistor R_1 and capacitor C_4 do not comprise a bias network as would be the case for a cathode resistor in vacuum tubes. Instead, this capacitor–resistor combination stabilizes operation of the transistor for temperature variations, as previously mentioned for the common-emitter amplifier.

Note the resemblance of the Class C amplifier to the TPTG oscillator shown in Fig. 3-10b. The presence of the input and output resonant circuits, plus the interelement capacitance, will tend to form an oscillatory system for

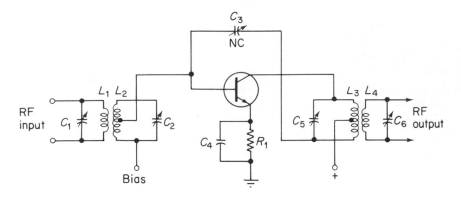

(a) transistor-type, series fed

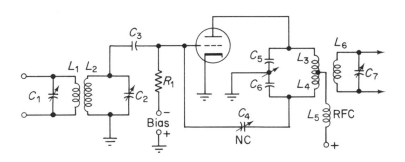

(b) tube-type, shunt fed

figure 4-12   class C amplifiers

the Class C amplifiers also. Thus, neutralization (unilateralization) is necessary and is accomplished by the neutralizing capacitors (NC) shown in Fig. 4-12. The output resonant circuit (*tank circuit*) is tuned by a single capacitor C_5. With such an arrangement, both the rotor and stator of the capacitor have a high potential present on them, and insulation precautions are essential in high-voltage operation. A better system is the split-stator process shown for the tube amplifier in Fig. 4-12b and discussed more fully later.

For the input system of Fig. 4-12b, a shunt-fed condition prevails, as opposed to the series-fed circuitry (Fig. 4-12a). Self-bias could be obtained for the Class C amplifier in Fig. 4-12b if we grounded resistor R_1. When the grid is driven positive for every other alternation of the input signal (see Fig. 4-2), the positive grid causes conduction of the positive pulses between the cathode and grid of the tube. Hence, a basic rectification process occurs for the input signal. This process charges C_3 with a polarity that makes the capacitor plate at the grid side negative, as opposed to the positive charge on the capacitor plate toward the resonant circuit. Consequently, the grid conduction lowers input resistance and a short time constant prevails. During the excursions of the excitation signal into the cutoff region, a long time constant prevails, since the grid is negative and does not conduct (high impedance). Thus, a charge builds up rapidly across C_3 (through the lower resistance path of L_2). Such self-bias is sufficient to drive the tube into the cutoff region for establishing cutoff bias. However, if excitation fails, no bias would develop, and tube currents would be excessive and could damage the system. Hence, the bias pack supplies the steady-state direct current and protects against excitation failure.

The output resonant circuit has the tapped inductance shunted by variable tuning capacitors C_5 and C_6. A single variable capacitor could be used, but the dual pair places the rotor section at ground potential and thus minimizes shock hazards. Since the rotors are connected to a common shaft and the stators are separated or split into two sections, these capacitors are often referred to as *split-stator* capacitors.

The supply potential is applied to the center of the anode inductance, forming L_3 and L_4 sections. Hence, both top and bottom of the resonant circut are above ground for neutralizing purposes.

If a pentode tube were used, the tendency to oscillate would be reduced, though not eliminated. At very high signal frequency operation, sufficient lead inductances and stray capacitances may be present to cause some parasitic oscillations, unless low-loss high-frequency tubes specifically designed for high-frequency operation are used, as was mentioned in Section 4-8.

For the circuit in Fig. 4-12a, the neutralization system could be called collector neutralization, since a portion of the RF energy is tapped from the collector inductance and fed back directly to the base. If the RF energy were tapped from the base inductor and coupled directly to the collector, it could be termed *base* neutralization. Similarly, in Fig. 4-12b plate neutralization is used. For grid neutralization it would be necessary to use a tapped L_2 inductance and to couple the RF energy directly to the plate.

The neutralization process consists of adjusting the neutralizing capacitor NC until the interelectrode or interelement capacitances are canceled. Initially, the supply voltage to the collector (or plate) is shunt off, and the antenna system or subsequent amplifier is uncoupled. An RF indicating device is used, such as an RF meter, or a dc ammeter with a diode rectifier in series with one lead. The indicating device is coupled to the output resonant circuit. With an RF signal applied to the input of the Class C amplifier, the output tuning capacitor (C_5 and C_6) is rotated through its full range to find a point of maximum needle deflection on the indicating meter. Now the neutralizing capacitor is adjusted for a minimum or zero reading. Recheck by again rotating the output tuning capacitor.

Another neutralization method consists of using a meter in the input (base or grid) circuit for reading grid current. The output tuning capacitor is again rotated, watching for a current change in the grid circuit. The neutralizing capacitor is adjusted until a rotation of the output tuning capacitor no longer causes a change of current reading at the input.

One Class C amplifier may be connected to another Class C stage, to a Class B linear, or to an antenna system. In all such instances the system to which the RF output is applied is called a *load* and, as such, draws power from the Class C stage feeding it. With no load applied, a minimum amount of anode (or collector) current flows, provided the circuit is tuned to resonance. At resonance, the parallel tank circuit has a very high impedance, and a maximum signal voltage develops across it. (Without a load, the signal amplitudes may become excessive because of the high impedance and may arc across the stator-rotor plates of the tuning capacitor.)

When the resonant circuit is tuned below or above resonance, excessive anode currents flow because the impedance drops to low values, as shown in Fig. 4-4. At resonance not only is the impedance high but also the circuit Q. Under load conditions, however, the Q drops to a value between 10 and 15 when the circuit is operating efficiently.

The desired Q under operational load conditions is obtained by proper design with regard to the ratio of L to C for the correct load resistance indicated by the tube or transistor characteristics. The low Q under load conditions is chosen to minimize the production of undesirable harmonics in the Class C amplifier. If, however, the Q is lowered excessively, the bandpass broadens and the rejection of undesirable harmonic signals is reduced. A lower-than-normal Q also reduces power output and circuit selectivity.

When the Q is too high under load conditions, anode or collector currents become excessive, and the resonant circuit itself tends to radiate signals to a greater extent than would normally prevail, resulting in RF power loss by improper radiation at the Class C amplifier instead of from the antenna.

The L and C values required to obtain resonance have been discussed in Section 3-3, and Q factors have been given in Section 4-3. As mentioned there, Eq. (4-10) ($Q = R/X_L$) indicates the resistance of the load applied to the resonant circuit. Such a load consumes power, and the load resistance is reflected to the resonant circuit (neglecting the very small inductor L_3 resistance). The value of the load resistance (R_L) is indicated by dividing the value of the applied dc anode (or collector) voltage by the amount of current flowing at the time the load is consuming power. Hence, by Ohm's law,

$$R_L = \frac{E_p}{I_p} \quad \left(\text{or } \frac{E_c}{I_c}\right) \tag{4-17}$$

Knowing the desired Q under load conditions, we can ascertain the capacitance value of a resonant circuit (for a given frequency). This capacitance is inversely proportional to the reflected R_L across the resonant circuit and, hence, inversely proportional to the E_p/I_p ratio. For the single capacitor shunting the inductor, as in Fig. 4-12a, the total capacitance is found by

$$C_T = \frac{300(QI_p)}{fE_p} \qquad (4\text{-}18)$$

where C_T = total capacitance in picofarads
  Q = a value between 10 to 15 under load conditions
  I_p = anode (or collector) current in milliamperes (dc)
  f = frequency in megahertz
  E_p = dc anode (or collector) voltage

Equation (4-18) provides the capacitance value for the approximate midtuning range of a variable capacitor. Knowing the capacitance value for the desired frequency at resonance and circuit Q under load, we can find the inductance value by Eq. (3-6).

4-10. split-stator factors

If split-stator capacitors are substituted for a single capacitor in the tank circuit of a Class-C amplifier, the coil size must be increased and the capacitor values decreased for the same power dissipation and operational characteristics. The reasons for such a change can be more readily understood by referring to Fig. 4-12 and the equivalencies shown in Fig. 4-13. In Fig. 4-12b, note that the cathode of the tube is grounded and, thus, common with the junction of the split-stator tuning capacitors C_5 and C_6. The anode, however, is connected to the junction of capacitor C_5 and the top of the L_3 part of the

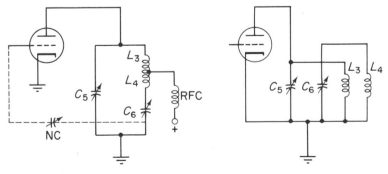

(a) equivalent circuit of 4-12(b)   (b) figure 4-13(a) redrawn

figure 4-13 split-stator equivalencies

tapped inductance. Thus, C_6 is actually in series with the inductance, as shown in the equivalent circuit of Fig. 4-13a.

Since the reactance of a capacitor electrically opposes the reactance of an inductor, the C_6 reactance (X_c) cancels some of the inductance reactance (X_L) of the coil. Hence, somewhere along the inductance, zero voltage prevails for the RF signal (a *node* point). Because such a zero-potential point is equivalent to an RF ground, the circuit is effectively split in two. (For Fig. 4-12b the nodal split would be across the ground junction of the capacitors and the tap on the coil.)

As the actual node point in the inductance may be difficult to localize accurately by the tap, it is left undisturbed. Thus, *no bypass capacitor* is attached at the tap, since it would place the point to which it is attached at RF ground. Because this point may not agree with the actual node point, circuit efficiency would be impaired and nodal equilibrium upset. Since the power supply is at RF ground (by virtue of the low-reactance filter capacitors), the tap for the supply voltage is also isolated from the power supply by the radio-frequency choke (RFC).

As shown in Fig. 4-13a, the anode of the tube is actually connected to only one half the resonant circuit. This is indicated by the redrawn version of Fig. 4-13b, where all actual or equivalent nodal points are shown at true RF ground. When the tube is across one half of the split-stator tank, *that half* should have the same Q under load conditions as a single-capacitor-coil combination.

With the split-stator arrangement, total current drawn from the supply system is still the same for a given power input to the tube. However, there is current division between the split portions of the circuit, since the section connected to the tube couples power into the split section (L_3 and L_4 can be considered *series aiding*). Thus, each portion now utilizes one half the total power. As the current divides but total power remains the same, it is evident that an equivalent parallel load resistance is now presented to the system. (The amplifier tube is considered as the RF generator in such discussions, since it is the source of RF power for the networks attached to it.)

Since an equivalent parallel system exists, each portion of the circuit must now have a load resistance equal to twice that of a conventional single-capacitor circuit, because cutting the current in half doubles the R_L value, as though we had increased the E/I ratio. With a doubling of the ratio (due to halving the current) and with C inversely proportional to the E/I ratio, C is cut in half for the portion of the tank circuit connected to the anode of the tube.

To bring this section of the circuit back into resonance again, the inductance must be doubled, since C has been halved. As the inductance in the second section of the circuit must also be doubled, the total inductance required will now be four times that which would be needed for the single-capacitor system. (Quadruple the inductance is obtained by doubling total inductor turns.)

Power output from the tank circuit is not increased when changing from a single capacitor to the split-stator arrangement, but only divides for each section of the resonant circuit. The inductor has a high-potential RF signal at each end, and current must still be supplied to each section. Total current remains unchanged, however, since the same *total* R_L is present. Even though the resistance reflected to each half of the inductance is twice that of the single-capacitor type, the two coil sections are equivalent to a parallel section, and the paralleling of resistances produces the same R_L as before.

Solving for one capacitance in the split-stator circuit on the basis of one half the circulating current in Eq. (4-18) indicates that the new value of the capacitance is one half that which would be used for a single-capacitor system. Thus, for split-stator operation, each capacitor is one half of what the single-capacity circuit would be. Total capacitance of the split-stator section is one quarter, and the total inductance of the entire coil (L_3 and L_4) is four times that of the single-capacitor circuit.

Once the total capacitance has been found for the required operating frequency of the Class C amplifier (with split stator), the value of the total inductance is ascertained by Eq. (3-6). Instead, the total inductance value for the split-stator arrangement can be found by

$$L_T = \frac{(159,223)^2}{Cf^2} \qquad (4\text{-}19)$$

where L_T = total inductance in microhenrys
$\quad C$ = capacitance in picofarads
$\quad f$ = frequency in kilohertz

4-11.   *push–pull class C*

In a push–pull circuit two transistors or tubes are used in a balanced arrangement. The system can be employed for any class of amplifier, and the Class C version using tubes is shown in Fig. 4-14. (Class A audio push–pull systems are covered in Chapter 5.)

Push–pull circuitry furnishes more than double the signal power output realized from a single tube or transistor amplifier. This comes about because of the reduction of harmonic distortion, thus salvaging signal energy. In *triodes*, second and even harmonics predominate, and the push–pull circuit minimizes such harmonics considerably. *Pentodes*, on the other hand, have harmonic distortion primarily of the *odd-harmonic* type, and push–pull circuitry is less effective in minimizing it. For any type of tubes or transistors, however, push–pull is an excellent system for obtaining a balanced circuitry

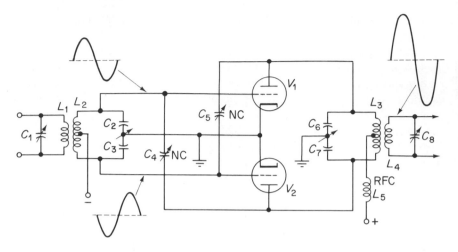

figure 4-14 push–pull class C

with increased power output. (Distortion factors are covered in greater detail in Chapter 5.)

In a push–pull system the input section of one tube or transistor must receive a signal 180° out of phase with the other. Thus, in Class B or C operation, the transistors or tubes conduct alternately, since one alternation of the input signal causes one tube to conduct and the other is driven into the cutoff region. As shown in Fig. 4-14, the RF excitation obtained from L_1 splits into a two-phase signal across L_2 because of the center tap (a nodal point for the RF). Thus, when V_1 grid has a positive alternation applied to it, it conducts, but at this time V_2 has a negative alternation present and, hence, this tube is driven farther into the nonconduction region. For the second alternation, V_2 conducts, but V_1 is cut off.

Thus, for push–pull Class C, we do not have to rely on the flywheel effect of the resonant circuit to supply missing alternations as in the case of single-ended RF Class B or Class C amplifiers (where the tube only conducts for alternate halves of the input signal). Instead, in push–pull the anodes conduct alternately; hence, one furnishes the positive output alternation and the other the negative. Consequently, more efficiency is realized, since power is delivered for each successive alternation of the output signal.

As shown in Fig. 4-14, the split-stator arrangement is used for both the input and output resonant circuits. The split-stator system is ideally suited to push–pull because of the symmetrical circuit division that exists. At the input, tube bias is applied at the center tap. For the output split-stator arrangement, the factors previously discussed apply. The rotors for C_6 and C_7 are at ground potential, thus minimizing shock hazards, while the power supply potential is applied to the tap at L_3 through the RFC (with no bypass capacitor used at the tap).

For push–pull, the inherent tendency of an RF amplifier to oscillate is eliminated by the unilateralization process known as *cross neutralization*. Capacitor C_5 connects from the anode of V_1 to the grid of V_2. By proper adjustment C_5 applies the necessary degenerative signal of proper proportion. (The phase of the output signal of V_1 is identical to that at the input of V_2.) Similarly, C_4 connects from the anode of V_2 to the grid of V_1. Thus, because of the phase reversals between the input and output of the tubes, and the phase differences at any instant between the individual grids as well as the individual plates, cross neutralization from the output of one tube to the input of the other is possible.

In practical transmitting systems, successive Class C stages between the oscillator and the final power amplifier bring the RF signals to the power requirements necessary for the desired transmission level. In commercial equipment, meters are placed in series with the grid and anode circuits for a constant reading of current amplitudes. An ammeter (or milliammeter) in the grid circuit aids in the grid tuning process and also permits adjustment of the amplitude of the excitation signal applied to the input sections. As a circuit is tuned to resonance there is a dip in the current reading (see Fig. 4-4). Similarly, a current meter in the anode circuit also aids in tuning. As described in Section 4-9, the grid meter also aids in the neutralization process.

4-12. frequency multipliers

In various broadcasting systems it is often necessary to increase the frequency of an RF signal. In particular, when very high frequencies are required for the carrier signal, it is often desirable to utilize a crystal of lower frequency for reasons of stability and ruggedness of the ground crystal. The required frequency is then realized by using Class C amplifiers for frequency multiplication.

Class C amplifiers can operate as frequency multipliers because they (as well as oscillators) operate on the nonlinear portion of the transistor's or tube's characteristic curve, hence generating a high order of harmonics related to the fundamental frequency.

By tuning the output resonant circuit to a harmonic of the fundamental frequency, this higher signal will develop across the resonant circuit. Plate (or collector) current will still be in pulse form but of a lower frequency. Thus, the output resonant circuit is supplied power less frequently than is the case when resonance is for the fundamental frequency. Consequently, efficiency and power output are less for the frequency doubler or multiplier.

A typical *frequency doubler* (or multiplier) is shown in Fig. 4-15. The base input circuit is tuned to 2 MHz, while the output collector section is tuned to twice this frequency (4 MHz). Note the absence of a neutralization capacitor. Since the resonant frequency in the collector side is different from that of the base side, no coinciding signal-frequency coupling exists and oscillations do not occur for the fundamental frequency. (For very high frequency operation, however, some oscillations could occur for frequencies above resonance owing to excessively long connecting leads, stray capacitances, and distributed capacitances in RFC units, which could make them behave as parallel-resonant circuits for some high frequency.)

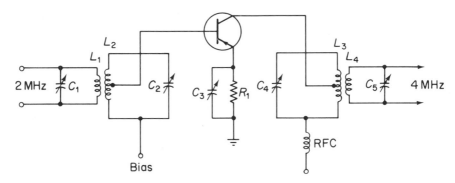

figure 4-15 frequency doubler

The amplifier of Fig. 4-15 can also act as a *tripler*, multiplying the input signal frequency by three. As such, however, efficiency is considerably lower than for a single-resonant Class C amplifier, because power is applied to the output resonant circuit only once for every three full sinewaves.

questions and problems

4-1. Describe the basic differences between an RF Class A and Class B amplifier.

4-2. Of what significance is the subscript 2 in AB_2?

4-3. Describe what is meant by *projected cutoff*.

4-4. What is the essential difference between Class B and Class C amplification?

4-5. Why is the efficiency of a Class C amplifier higher than the Class B type?

4-6. Compare the current and impedance changes in a series-resonant circuit with that of a parallel-resonant circuit as each is tuned through resonance.

4-7. Explain what is meant by the *half-power* point frequencies.

4-8. What circuit changes must be made to alter the shape and bandpass of a circuit's resonant curve?

4-9. If the Q of a resonant circuit is 200 and the resonant frequency (f_r) is 2 MHz, what is the bandwidth?

4-10. A series-resonant circuit has a resistance of 25 Ω an inductance of 200 μH, and a capacitance of 0.0002 μF. Determine the Q, the bandwidth, and the resonant frequency.

4-11. For the circuit in Problem 4-10, prove Q by using the equation $Q = X_L/R$.

4-12. What occurs to the frequency and Q if we double the inductance value and halve the capacitance of a series-resonant circuit?

4-13. What advantages, if any, does the common-base amplifier have over the common-emitter type?

4-14. Describe what is meant by the terms *unilateralization* and *neutralization*.

4-15. In circuitry design, what is meant by *shunt feed* and *series feed*?

4-16. Describe the differences between loose coupling, critical coupling, and tight coupling.

4-17. What term describes the percentage of coupling and what equation solves for it?

4-18. In an RF amplifier system two resonant circuits are coupled by a transformer arrangement between the inductances. The primary has an inductance of 12 μH, and the secondary, 3 μH. The k is 0.75. What is M?

4-19. When several RF amplifier stages are present, what two methods can be used to increase bandpass?

4-20. What type of RF amplifier is used to increase the power of a modulated carrier? Explain your designation.

4-21. In what manner does a Class C amplifier resemble a TPTG oscillator?

4-22. Define the terms *excitation* and *tank circuit*.

4-23. Describe *split-stator* tuning.

4-24. What are the disadvantages of too low or too high a circuit Q?

4-25. What load resistance is indicated in a Class C amplifier if $E_p = 2000$ and and $I_p = 0.5$ A?

4-26. Why is it preferable not to attach a bypass capacitor at the point where the supply voltage taps the resonant circuit inductance in a split-stator system?

4-27. For split-stator operation, what should be the value of each capacitance compared to that used for a single-capacitor system?

4-28. What are the advantages of push–pull operation for Class C amplifiers?

4-29. Describe *cross neutralization*.

4-30. Explain briefly why a Class C amplifier is capable of operating as a frequency multiplier.

5

AF and video amplification

5-1. introduction

When a transducer, such as a microphone, converts an audible signal into an electric counterpart, the output amplitude is too feeble for practical use. Hence, some form of amplification must be employed to raise signal levels to that required for loudspeaker operation. Similarly, the output from a radio detector (which converts the modulated portion of an RF signal into an audio-signal counterpart) must also be amplified an additional amount, as shown in Fig. 5-1a. The output from other transducers, such as tape playback heads, phonograph cartridges, etc., all require one or more stages of audio amplification before the audio signal has attained usable amplitude.

As shown in Fig. 5-1a, when very low level signals are encountered, the stage furnishing preliminary amplification is termed a *preamplifier*. A stage accepting a medium-level signal, such as from a preamplifier or radio detetector, or from a tape deck already having preamplification, is called a *voltage amplifier*, since no power is utilized for application to a speaker. The final stage is called a *power amplifier*, since both signal voltages and currents are needed to operate the voice coil of a loudspeaker.

When an audio signal is used to modulate an RF carrier, the final audio-amplifier stage is called a *modulator* instead of a power amplifier, though it still handles audio power and the circuitry is almost identical to the system shown in Fig. 5-1a. The sequence of amplification stages is also similar (see Fig. 5-1b). The essential difference is the output coupling employed, as more fully described in Chapter 6.

The frequency range of a high-fidelity audio-amplification system may

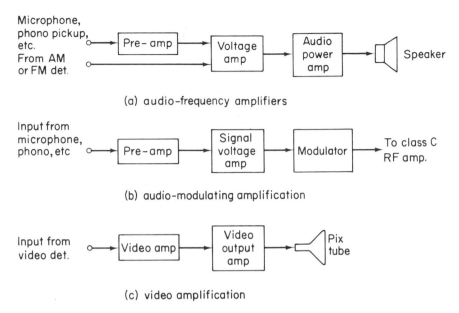

(a) audio-frequency amplifiers

(b) audio-modulating amplification

(c) video amplification

figure 5-1 AF and video amplifier applications

extend from 30 Hz to well over 20 kHz, though the actual range in disk and tape recordings may be somewhat less, depending on processing techniques. The video signal in television has a wide frequency range of from 30 Hz to 4 MHz. Though the low-frequency end is in the audio region, the upper frequency actually extends into the RF region (4000 kHz), and special wide-band amplifiers are necessary to preserve signal fidelity during amplification. In television receivers, one or two stages of video amplification are usually sufficient, as shown in Fig. 5-1c.

The circuitry involved in audio and video amplification is covered in this chapter. Modulation and detection factors for audio and video signals are covered in Chapters 6 through 9. The characteristics of audio signals were covered in Chapter 1, including pulse factors applicable to video signals. Additional details on video-signal composition are given in Chapter 8.

5-2. coupling methods

As discussed in Chapter 4, the type of amplification encountered in practical applications ranges from Class A through Class C. For audio amplification when signal *voltages* only are involved, Class A is used. For power amplification, Class A, AB_1, or AB_2 is used. Class C is not practical for use as audio power amplification. Power amplifiers are used to drive loudspeakers, or other power-amplifier stages that consume signal power at their input. The modulator stages of communication systems are also power amplifiers.

An important aspect of audio circuitry is the method used to interconnect successive amplifier stages. A commonly used system is shown in Fig. 5-2a,

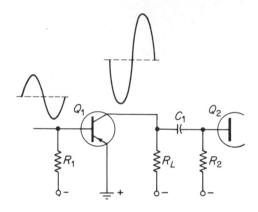

(a) RC coupling

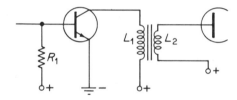

(b) transformer coupling

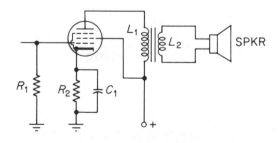

(c) output transformer coupling

figure 5-2 RC and transformer coupling

and this type of interstage coupling is known as *resistance–capacitance* coupling. As shown, the amplified signal energy appearing at the output of transistor Q_1 develops across the load resistor R_L and is coupled to the base of Q_2 by virtue of the low reactance of the series coupling capacitor. Infrequently, an iron-core inductor may be used instead of the resistor R_L, though the bulk and cost of the inductor for low-frequency audio signals discourage its general use. For RF the inductor has less turns and a smaller core, and, hence, finds greater use when a high impedance is required in the coupling system.

The series coupling capacitor has a lower reactance for higher frequencies; hence, high-frequency signals at the input of Q_2 have a higher amplitude than low-frequency signals (assuming all have the same amplitude at the output of Q_1). The coupling capacitor C_1 and input resistor R_2 comprise a series circuit across R_L. Thus, if R_2 has a value of 5000 Ω and C_1 has a reactance value that is also 5000 Ω for a specific signal, only half the voltage is applied to the base of Q_2, since the base taps the junction of C_1 and R_2. If the reactance of C_1 drops to a lower value for a higher signal frequency, more energy is coupled to the base input of Q_2.

In resistance–capacitance coupling the interstage capacitor values are selected to provide a maximum signal transfer consistent with physical size, voltage rating, and cost factors. For tube circuits higher voltage ratings are necessary, but because of higher input impedances and higher R_2 values, capacitance values range from approximately 0.01 to 0.5 μF. Too large a value is avoided because the bulk may present some capacitance to ground, resulting in some higher-frequency signal attenuation. For transistor circuits in which input and output impedances are much lower than tube circuits, the interstage capacitors have much higher values (though smaller because of lower voltage ratings). Approximate values range from 0.1 μF to well over 20 μF.

In Fig. 5-2b is shown transformer coupling between stages. This system has the advantage of matching the impedances between output and input stages, or, as shown in Fig. 5-2c, between the output of an amplifier and a speaker. The ratio of primary to secondary turns of wire in the transformer determines the impedance matching characteristics:

$$\text{turns ratio:} \quad \sqrt{\frac{Z_1}{Z_2}} \tag{5-1}$$

Thus, if an interstage transformer is to match an output impedance of 45,000 Ω to an input impedance of 5000 Ω, the turns ratio between primary and secondary should be 3 to 1:

$$\sqrt{\frac{45,000}{5000}} = \sqrt{9} = 3$$

For an output transformer the same procedures are followed to find the turns ratio. Assume that the output impedance of 6400 Ω is to be matched with a 4-Ω speaker voice coil:

$$\sqrt{\frac{6400}{4}} = \sqrt{1600} = 40$$

Thus, the output transformer should have a turns ratio of 40 to 1. These calculations do not indicate the exact number of turns, only the ratio of turns. The exact number of turns depends on the amount of core material, the type of core, and the efficiency of the winding. Hence, for a 40:1 ratio the primary could have 400 turns and the secondary 10 turns. The impedances would still match if the primary had 600 turns as against 15 for the secondary winding.

As with the *RC* coupling method, the transformer coupling introduces some losses. Since the inductor shunts the signal and has a decreasing reactance for lower signal frequencies, an uneven response results. Thus, if the input signals are constant in amplitude, the output signals would have lower amplitude as their frequencies dropped. In addition, the distributed capacitance between coil turns and layers offers a decreasing shunt reactance for the higher frequencies. The combined effects of coil and capacitor reactances limit the extent of the even output response because of the attenuation of the upper and lower signal frequencies.

Good transformer design extends the response and reduces losses within certain limits. With an increase in core size the number of turns (though not the ratio of turns) can be decreased, reducing the effects of distributed capacitance and shunt inductive reactance. Thinner wire also reduces capacitance effects, though the increased wire resistance introduces losses for all frequencies and wastes signal energy. Because of the increased costs and bulk of high-quality transformers, their use is avoided when possible in favor of other coupling methods.

The disadvantages of *RC* and transformer coupling are eliminated by using *direct coupling* (DC). As shown in Fig. 5-3, there is a direct path for direct current as well as alternating current between the output of one stage (Q_1) and the other (Q_2). Because of the impedance mismatch between stages and the voltage differences between the output and input circuits, design factors are more critical than usual. The mismatch may necessitate the inclusion of an additional amplification stage or the use of higher-gain transistors (or tubes).

While similar transistors can be used in direct coupling, the circuit in Fig. 5-3 uses dissimilar (complementing) transistors to simplify voltage distributions. The first transistor (Q_1) is an *npn* type and obtains the required negative potential for the emitter from the common ground circuit. Note that the positive potential is applied to the top of R_2, which in conjunction with R_3 acts as a voltage divider (R_2 and R_3 shunt the battery). The ratio of

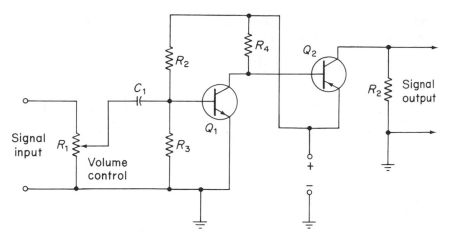

figure 5-3 dc amplifier (complementing transistors)

the individual resistor values is selected to apply the proper potential to the base of Q_1. If, for instance, the voltage source is 9 V and 3 V is required at the base, the two resistors would have equal values, such as 33,000 Ω each.

The signal voltage develops across R_4, and this resistor also couples the necessary positive supply voltage to the collector for reverse bias. The positive potential is also applied to the emitter of Q_2 for the proper polarity forward bias for this *pnp* type. The required negative potential for the collector of Q_2 is obtained from the ground circuit via the output resistor R_5.

For Q_2 the forward bias between base and emitter requires a negative potential at the bias. While the base of Q_2 in Fig. 5-3 may seem to be positive, because of its connection to the positive terminal of the source potential, circuit design satisfies the bias requirements. This comes about because resistor R_4 is in series with the impedance existing between the collector and emitter of Q_1. Hence, R_4 and the transistor impedance also shunt the applied source potential, thus forming a voltage divider. Thus, the junction of R_4 and the collector of Q_1 forms a negative terminal with respect to the positive terminal of the source potential. Hence, the base of Q_2 is negative with respect to the positive-potential emitter for the required forward bias between base and emitter.

5-3. attenuators and pads

Resistive networks are extensively used to diminish signal strength to that required, or to match dissimilar impedances (or both). The terms *attenuator* and *pad* have been used synonymously for such circuits. The *tone control* is an attenuator without serving as an impedance matching unit. Such tone controls are used in audio systems to alter the response curve by diminishing either the high- or low-frequency signals as desired. This is done at a sacrifice in overall gain, because we diminish lower-frequency signals and thus, by contrast, raise the high-frequency signals. Similarly, if we attenuate the high-frequency signals, the lower-frequency signals become louder (in comparison to the diminished highs). Thus, after diminishing certain signals, it is necessary to readjust the gain control to achieve the same output level as before.

A typical bass tone control is shown in Fig. 5-4a and consists of the

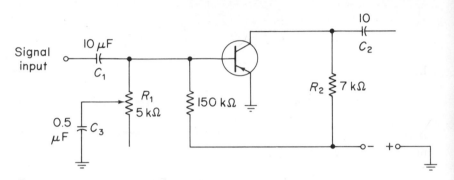

(a) bass tone control

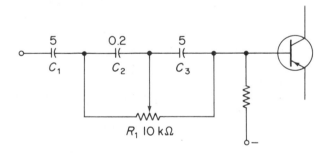

(b) treble tone control

figure 5-4 tone controls

added components R_1 and C_3. With the arm of the variable resistor R_1 at the upper position, the signal components appearing at the base input encounter a shunting effect, because C_3 now parallels the base input. Hence, signals having higher frequencies are attenuated and the bass tones are, by contrast, accented. Similarly, if the variable arm of R_1 is at the lower end, the resistance of R_1 minimizes the signal-shunting effect of the capacitor C_3 and decreases (by amplitude comparison) the bass response. Intermediate settings of R_1 produce various degrees of control.

A treble tone control is shown in Fig. 5-4b. When the variable arm of R_1 is positioned at the left, it effectively shorts out C_2, permitting the lower reactance of C_3 (because of its larger value) to pass all signals at normal attenuation. When the variable arm is at the right, however, C_3 is shorted, leaving C_2 in series with the signal input. The higher reactance (because of the lower capacitance value) provides an increasing attenuation for signals of lower frequencies. Thus, since the lower-frequency signals are attenuated, the higher-frequency signals (by comparison) appear to have been increased in amplitude. Again, intermediate settings of R_1 provide variations between maximum and minimum changes in the treble response.

The volume control shown in Fig. 5-3, as with the tone controls, is an attenuator without impedance matching. By adding one or more resistors to the gain-control section, impedance matching can also be achieved. On occasion the purpose for using pads may be for impedance matching alone without signal attenuation being desired. Since, however, a resistive network is involved, the slight loss of signal strength is tolerated to accomplish the primary objective. Since communications circuitry may be operating below the maximum possible signal amplitude that could be obtained by the amplification circuitry, the gain can be increased to compensate for the amplitude loss introduced by impedance-matching pads.

The single potentiometer gain or tone control presents a varying impedance as the variable arm is rotated. By using an additional resistor to form what is known as an L pad, impedance matching can be maintained. Some pads are fixed and may be of two types: the asymmetrical and the symmetrical. The asymmetrical types are those having dissimilar impedances at the input and output. Such pads are used to match the output impedance of one circuit to that of the next circuit. The symmetrical pads have identical input and output impedances and their purpose is to introduce a desired amount of attenuation to bring the signal levels down to those required. Both the symmetrical and asymmetrical pads may be of the unbalanced type (one line grounded and the other above ground) or the balanced type (both lines above ground).

The basic L-type fixed pad is shown in Fig. 5-5a, which illustrates the asymmetrical pad serving to match Z_1 to Z_2, that is, to match the impedance of an input circuit to that of the circuit connected to the output of the pad. This L-type pad is a *minimum-loss type* since it only introduces a minimum amount of loss while matching the two impedances.

The L pad can step up the impedance or decrease it as required. If Z_1, for instance, is 300 Ω and is to be matched to the impedance of a subsequent circuit (Z_2) of 50 Ω, the network shown in Fig. 5-5b would be used. Note that the Z_1 section of 300 Ω "sees" a resistor of 280 Ω in series with a parallel branch of R_1 (56 Ω) and Z_2 (50 Ω). By Ohm's law, the total resistance value that Z_1 then sees is sufficiently close to its own 300-Ω value to provide a satisfactory impedance match. The output device represented by Z_2 sees a

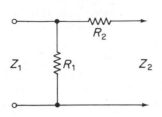

(a) basic L-type fixed pad

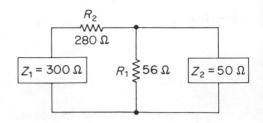

(b) typical fixed L-pad values

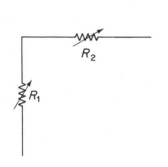

(c) variable L-pad

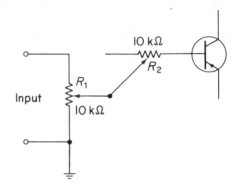

(d) typical values for variable L-pad

figure 5-5 fixed and variable L pads

shunting resistor of 56 Ω, which is paralleled by a series branch of 280 and 300 Ω. This produces a value of approximately 50 Ω, or one sufficiently close to this value when standard resistor values are employed.

The resistors and impedances have the following relationship:

$$R_1 R_2 = Z_1 Z_2 \qquad (5\text{-}2)$$

If Z_1 has a lower impedance than Z_2, we may express the relationship as follows for whole-number convenience in the ratio of the impedances:

$$\frac{R_2}{R_1} = \frac{Z_2}{Z_1} - 1 \qquad (5\text{-}3)$$

If the value of R_1 is known, we can solve for R_2 using

$$R_2 = \left(\frac{Z_2}{Z_1} - 1\right) R_1 \qquad (5\text{-}4)$$

Thus, for Fig. 5-5b, we solve for R_2 using Eq. (5-4):

$$R_2 = \left(\frac{300}{50} - 1\right) 56 = 5 \times 56 = 280 \ \Omega$$

If both resistor values are unknown, their individual values can be found by

$$R_1 = \frac{Z_1}{\sqrt{1 - Z_1/Z_2}}$$

$$R_2 = Z_2\sqrt{1 - \frac{Z_1}{Z_2}}$$

(5-5)

After the calculations have been performed, standard resistor values are selected that are closest ot the calculated values.

In Fig. 5-5c is shown the conventional representation for a variable L pad. In practical applications two variable resistors are hooked up to form the network shown in Fig. 5-5d. With such an arrangement the value of the shunt resistance (10,000 Ω) is maintained for the input of the transistor, regardless of the setting of the gain control. For maximum signal application to the base of the transistor, the variable arm for R_1 is at the top and that for R_2 at the extreme right position.

Under these conditions only the resistance of R_1 (plus the impedance of the input device) shunts the input of the transistor. For other volume settings, the same shunting impedance and resistance prevail. For the half-volume setting, for instance, each variable arm taps half the resistance (assuming a linear tapper in the resistance strip). Thus, the input to the transistor is still shunted by 10,000 Ω (5000 Ω for R_1 and 5000 Ω for R_2 now in series). Now, however, a 10,000-Ω resistance prevails between the input of the transistor and the signal source. At the minimum volume setting the R_1 arm would be at ground and the arm for R_2 at the extreme left. Now no signal is applied to the base of the transistor because of the grounding of the R_1 variable arm. Resistor R_2 now applies its full 10,000 Ω between the transistor base and ground, maintaining the same shunt resistance.

A balanced arrangement of the L pad is shown in Fig. 5-6a. Since it resembles the letter U on its side, it has been referred to as a U pad also. Note that each series resistor is one half the value of R_2 shown originally in Fig. 5-5a. If, at the input, an impedance (Z_1) of 50 Ω is to be matched with an impedance Z_2 of 300 Ω, the values for the resistors are as shown in Fig. 5-6b. Compare these values with those shown in Fig. 5-5b.

Attenuators and pads placed in cascade (see Fig. 5-5c) form a *ladder* pad. As *more* resistors are added to the ladder arrangement, the attenuation factor *increases*.

Variable and fixed T pads, plus a symmetrical H pad, are shown in Fig. 5-7. The standard symbol for the variable T pad is shown in Fig. 5-7a. Practical application of this pad is as shown in Fig. 5-7b, where three variable resistors are so arranged that the arm for R_1 is at the left, the arm for R_2 at the right, and the arm for R_3 at the top for a maximum transfer of signal between the input device and the base of the transistor. At zero input the R_1 and R_2 arms are at the open ends of the resistors, and the arm for R_3 is at

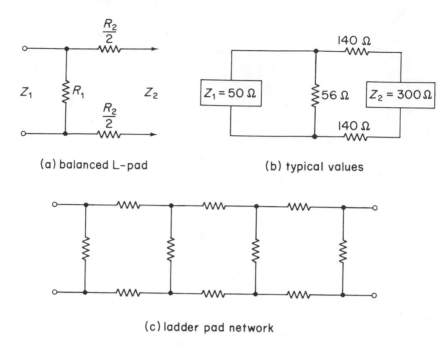

(a) balanced L-pad

(b) typical values

(c) ladder pad network

figure 5-6 balanced L and ladder pads

ground. Again, a constant shunt impedance is maintained between the base and emitter input of the transistor.

A fixed symmetrical T pad is shown in Fig. 5-7c, where the impedance of the input device matches that of the output device and attenuation is the sole intended function, without introduction of an impedance mismatch. Since there is no need for matching impedances, all R_1 resistor values are identical, with R_1 and R_2 selected to introduce the degree of signal attenuation required. The T pad (Fig. 5-7c) is an *unbalanced* type, with the balanced equivalent shown in Fig. 5-7d, where the R_1 values must be halved in comparison to the values used in Fig. 5-7c.

To ascertain the R_1 and R_2 resistance values, we must relate them with the ratio of signal-voltage or signal-current attenuation needed. Hence, the equations for solving for R_1 and R_2 must incorporate the ratio of attenuation desired between the input and output of the pad. Either voltage or current could be used, but for purposes of illustration we shall use the voltage ratio designated as V:

$$R_1 = Z\left(\frac{V-1}{V+1}\right) \tag{5-6}$$

$$R_2 = Z\left[\frac{2V}{(V+1)(V-1)}\right] \tag{5-7}$$

Thus, for the T pad of Fig. 5-7c, if the signal voltage has an amplitude

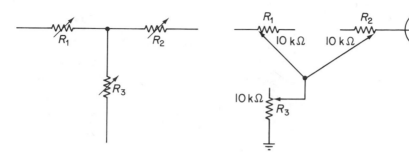

(a) variable T-pad   (b) typical values and application

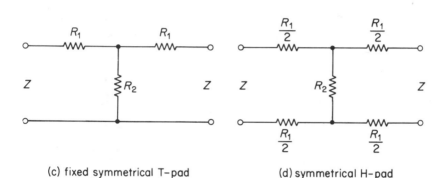

(c) fixed symmetrical T-pad   (d) symmetrical H-pad
(unbalanced type)

figure 5-7 variable and fixed T pads, plus symmetrical H pad

of 100 V and is to be attenuated to furnish a 10-V output signal, the voltage ratio (V) would be 10. If the impedance is 75 Ω, the value of R_1 is

$$R_1 = 75\left(\frac{10-1}{10+1}\right) = 75\left(\frac{9}{11}\right) = 75 \times 0.8 = 60 \ \Omega$$

Figure 5-8a shows the *pi* pad, which is a symmetrical type, though unbalanced with respect to ground. The balanced version (Fig. 5-8b) is referred to as the O pad because of the similarity in configuration between it and the letter O. Because impedances are similar, no matching design is involved, and the resistance values are selected to introduce the desired amount of signal attenuation. As with the T pads, the equations for solving the resistor values must be based on the voltage or current attenuation ratios. The voltage-attenuation ratio will be designated by

$$R_1 = Z\left(\frac{V+1}{V-1}\right) \tag{5-8}$$

$$R_2 = Z\left(\frac{V^2-1}{2V}\right) \tag{5-9}$$

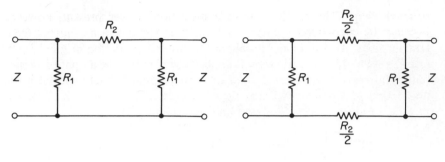

(a) Π–pad, symmetrical type, unbalanced   (b) O–pad, symmetrical type, balanced

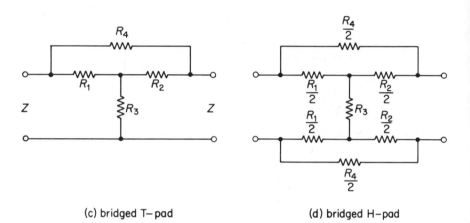

(c) bridged T–pad   (d) bridged H–pad

figure 5-8 pi and O pads, plus bridged T and H pads

An added resistor may be shunted across the series resistors of T and H pads to form what is termed a *bridged pad*. The bridged T and H pads are shown in Fig. 5-8c, and d with the R_1 and R_2 values equaling that of the impedance. Because the ohmic values of R_1 and R_2 individually must match the Z, only resistors R_3 and R_4 need be calculated. The formulas used are

$$R_3 = \frac{Z}{V - 1} \tag{5-10}$$

$$R_4 = Z(V - 1) \tag{5-11}$$

5-4.   signal distortion

Since the purpose of an amplifier is to increase signal levels, it should do this without adding additional signals to the original or distorting the original signal. The tendency toward distortion is ever present, however, because the characteristic curves of transistors (or tubes) are not perfectly straight lines. This nonlinearity alters the waveshape of the original signal (see Fig. 5-9a). If a pure sinewave is applied to the input, the amplified version (either in terms of collector current or output signal voltage) will have one alternation greater in amplitude then the other. This *amplitude distortion*, in turn, produces *harmonic distortion* (see Section 1-4).

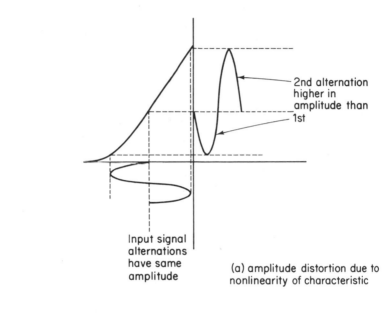

2nd alternation higher in amplitude than 1st

Input signal alternations have same amplitude

(a) amplitude distortion due to nonlinearity of characteristic

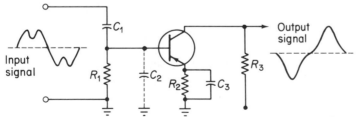

(b) phase distortion from input to output

figure 5-9 harmonic distortion and phase distortion

The type and degree of distortion depend on various factors, including the characteristics of the transistor or tube involved. Triodes, for instance, produce a distortion primarily of the even-harmonic type. The second-harmonic signal is the most prominant, with higher orders of even harmonics progressively lower in signal amplitude. Pentode tubes produce a harmonic distortion predominantly of the odd-harmonic variety, with the third harmonic highest in amplitude.

The class of amplifier also determines the degree of distortion, with Class A producing the least amplitude change of one alternation with respect to the other because operation is over the most linear portion of the characteristic curve. (See Fig. 4-2 and Section 4-2). In high-quality amplifiers the design holds harmonic distortion to a minimum, often less than 1 per cent. When amplification only involves speech reproduction or when quality of music is not too important a factor, distortion rates as high as 5 to 10 per cent may be encountered. Distortion below 5 per cent is usually not noticeable to the average ear.

In Section 5-2 factors of distributed capacitance, series reactances, etc., were discussed. Attenuation for certain signals of specific frequencies results in an uneven amplifier response and is known as *frequency distortion*. Interelement capacitances in transistors (and interelectrode capacitances in vacuum tubes) also have a low shunt reactance for higher-frequency signals, contribute to frequency distortion, and prevent a flat output response.

As mentioned in Section 5-2, a coupling capacitor plus the input resistor form a voltage divider for the signals applied to the input. This circuitry is shown in Fig. 5-9b. Actually the base-emitter input is not purely resistive since R_1 is shunted by C_2, which represents the capacitance between the base and emitter (or between the grid and cathode of a tube). Thus, R_1 and C_2 form an impednace that has a lower ohmic value for higher-frequency signals, contributing to an uneven response. At lower-frequency signals the reactance of C_1 increases and a larger signal voltage appears across it, with a corresponding decrease of signal voltage across R_1, which feeds the input to the transistors.

The changing reactances of C_1 for signals of varying frequencies, plus the shunting interelement capacitances of the transistor, causes a phase shift between high- and low-frequency signals in addition to the frequency distortion encountered. When higher-frequency signals (or the high-frequency signal components of a complex waveform) undergo such phase shifting with respect to the lower-frequency signals, *phase distortion* occurs. This is shown in Fig. 5-9b, where the input signal consists of a fundamental plus a third-harmonic component. The output waveform still contains the two signals, but the shift in phase between the fundamental and third harmonic results in the phase distortion of the amplified signal in comparison to the original waveshape at the input. This type of distortion can have serious consequences in amplifiers handling visual-type signals, such as in television, radar, or in oscilloscopes, where the shift in picture information causes image blurring.

Distortion is minimized by careful design practices, including use of direct coupling (DC) whenever possible, selection of transistors or tubes having low interelement capacitances, keeping leads short (to minimize lead inductances at higher frequencies), and also spacing between connecting wires to minimize stray capacitances. Push–pull circuitry can be used in the output-power stages as an additional aid toward the reduction of distortion.

In push–pull, two transistors or tubes are used in a circuit in such a manner that the signals applied to the individual inputs are out of phase with each other, as shown in Fig. 5-10. Hence, at any instant, the input signal for

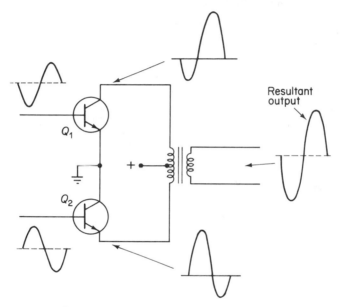

figure 5-10 push–pull input-output signals

one transistor adds to the forward bias and increases collector–emitter conduction, while the signal at the input of the other transistor reduces forward bias and decreases conduction. In consequence, electron flow from emitter to collector increases for one transistor and decreases in the other. The combined voltage drop across the primary halves (and the combined fields produced by the currents) provides a secondary output wherein amplitude distortion has been minimized.

How such distortion is reduced is illustrated in Fig. 5-11. Assume that the decrease in collector current (representative of a negative signal-current change) does not reach as high an amplitude as a positive change, owing to the nonlinearity of the characteristic curve. When one transistor has a negative change, however, the other has a positive change, and the resultant peak-to-peak currents produce a virtually identical amplitude for each alternation at the output from the transformer secondary.

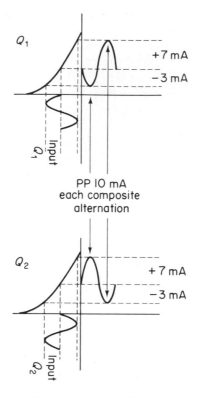

figure 5-11 distortion reduction in push–pull

The opposing current changes in the push–pull transformer also have the effect of reducing core saturation. With triodes in push–pull operation there is considerable reduction in even-harmonic distortion. Distortion robs the amplifier of some of the power output of the distorted signal, since power output at any instant is determined by the applied voltage, transistor characteristics, and input signal amplitudes. Thus, the signal-power output from a triode push–pull amplifier will be more than double the power developed by a single transistor or tube, because the undistorted output increases.

Pentode tubes produce a high degree of odd-harmonic distortion and, since push–pull favors reduction of even harmonics, pentode use is not as advantageous regarding distortion reduction. Pentode and beam-power tubes are found in some audio-modulator applications, however, where their particular advantages are useful in special applications.

For proper operation of push–pull circuitry, a balanced system is necessary. Components for each half of the push–pull section should be well matched, and the transformer should be well designed so the center tap is as close to producing an equal division of impedances as possible.

5-5.   *push–pull audio circuitry*

One method for obtaining the necessary out-of-phase signal voltages for the inputs of the audio push–pull system is shown in Fig. 5-12a. A transistor could also be used, with one output from the collector and the other from the

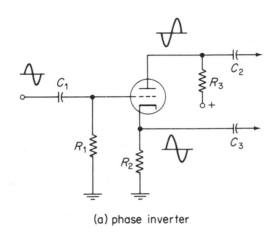

(a) phase inverter

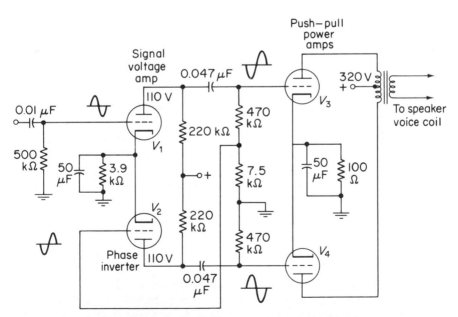

(b) symmetrical phase inversion push–pull amplifier

figure 5-12 tube-type push–pull audio amplifier

emitter. For the circuit shown, there is the normal phase inversion between the input signal and the amplified version appearing at the plate. At the cathode, however, the signal is in phase with that at the input. The output from across R_2 is known as a *cathode follower* (the output phase *follows* that of the input), while for a comparable transistor version, the circuit is referred to as an *emitter follower*.

A more symmetrical phase-inversion system is shown in Fig. 5-12b. The additional tube also increases efficiency and provides an independent output signal, as shown. The input signal for the grid of the phase inverter V_2 is obtained from a signal-voltage tap in the grid-resistor network at the input of V_3. The gain of V_1 determines the value of the resistor between the tap and ground. If, for instance, the gain is 60, the tap provides one sixtieth of the amplified voltage and applies this to the grid of V_2, so it has the same amplitude input signal as V_1.

For optimum performance, a high-efficiency output transformer must be used, and tubes (or transistors) V_1 and V_2 should be closely matched, as should also V_3 and V_4. The plate-to-plate impedance of the push–pull tubes is matched to the ohmic value of the voice-coil impedance by the proper turns ratio of the output transformer, using Eq. (5-1).

A transistorized push–pull audio amplifier that dispenses with the output transformer is shown in Fig. 5-13a. An interstage split-secondary transformer is used to furnish the necessary out-of-phase signal voltages to the base inputs. Resistors R_1 and R_2 shunt the voltage source E_1, and the necessary forward bias for transistor Q_2 is obtained from across R_2, making the base negative (via the transformer secondary) and the emitter positive, as required for the *pnp* transistor. For *npn* types the supply potentials would be reversed.

For Q_3, resistors R_3 and R_4 shunt E_2 and forward bias is obtained from R_4. In this circuit the speaker inductor is in series with one leg of each power source. Because of the low impedances in transistor circuitry, resistor values are also low. With 2N2147 transistors, for instance, typical values at the output for Q_2 and Q_3, would be: R_1 and R_3, each 330 Ω; R_2 and R_4, 3.9 Ω; and the stabilizing transistors R_5 and R_6, 0.27 Ω each. For these transistors each voltage source would be approximately 22 V and the speaker voice-coil impedance 4 Ω.

Note that resistors R_1 through R_4 are in series across the two power sources, as are the collector-emitter sections of Q_2 and Q_3. Circuit design, however, is such that the output transistors are in parallel with the load (the speaker voice coil). This is evident in Fig. 5-13b, which shows the equivalent circuit. With well matched transistors each will have a voltage drop across it equal to the other. With both power sources the same the network bridge is balanced and no direct current flows through the speaker inductor. When the out-of-phase signals appear at the inputs to the transistors Q_2 and Q_3, one transistor conducts more than the other, upsetting the equality of currents in the bridge circuit. Now audio-signal voltages appear across the speaker voice-

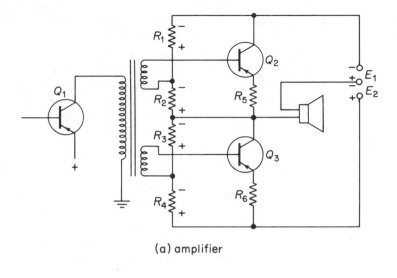

(a) amplifier

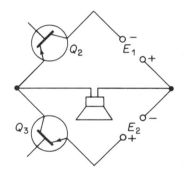

(b) equivalent circuit

figure 5-13 transistor push–pull audio amplifier and equivalent circuit

coil inductor; hence, audio-signal currents flow through the speaker, producing sound output.

In vacuum-tube power-output audio stages, the high plate impedance
requires a step-down transformer for matching impedances to the low-ohm-
speaker voice coil. With the circuit shown in Fig. 5-13, however, a good match
is obtained because of the low impedance of the transistor circuitry.

5-6. inverse feedback

An additional reduction in distortion can be achieved by using *inverse
feedback*. This system also reduces circuit noises, though there is a reduction
of signal gain. Inverse feedback can be of the signal-voltage or signal-current

type, and both are used simultaneously on occasion. Inverse (negative) signal-voltage feedback applies a portion of the amplified output signal to an earlier circuit in such a manner that the signal fed back is out of phase with the existing signal at the place to which the feedback signal is applied.

Inverse feedback is shown in Fig. 5-14a for a transistor audio amplifier similar to that of Fig. 5-13. For Fig. 5-14, Q_3 emitter resistor is grounded, as is one leg of the speaker inductance. A 200-μF coupling capacitor isolates the dc potentials of the two stages. The voltage drop across the speaker impedance serves as the source voltage for feedback purposes. As shown, the coupling is back to the emitter of a previous stage and develops across the lower resistor. The upper network consisting of the 1000-Ω resistor and shunting 100-μF capacitor is for current stabilization for thermal changes that might affect transistor conduction, and is unrelated to the feedback loop. (If a temperature change causes an increase in transistor current, a larger drop appears across the 1000-Ω resistor, and the resultant bias change tends to reduce current and thus to stabilize the system.)

The amplitude of the feedback voltage can be regulated by changing the values of the series resistor and the emitter resistor. Since the feedback signal appearing across the emitter resistor is out of phase with the incoming signal developed in this circuit, some degeneration occurs, and the amplifier output signal is reduced in proportion to the amount of voltage fed back.

The signal fed back, however, contains a portion of the distortion that appeared at the output. This distortion signal is also coupled to the emitter circuit and is, in turn, amplified in the same manner as the original input signal. The amplified distortion component is out of phase with the distortion developed within the transistor; hence, cancellation occurs to the degree of the amplitude relationships between the internal distortion and that fed back.

With a transformer output, the feedback voltage can be obtained from the secondary winding, as shown in Fig. 5-14b. This could be a transistor circuit instead of the tube shown. If the fed-back voltage is found to be in phase instead of out of phase, the ground terminal and the feedback tap at the secondary are reversed. The amplitude of the feedback voltage is determined by the values of the series resistor plus the cathode resistor. The series capacitor in the feedback loop blocks the direct current developed across the cathode resistor from the transformer secondary.

In the design of feedback systems, the lowercase Greek beta (β) is used to indicate feedback-voltage *amplitude*. A minus sign precedes the letter when the feedback signal voltage is of an inverse or negative nature. Strictly, β indicates the *decimal equivalent of the percentage of output signal voltage that is fed back*. Since the amplification is altered by inverse feedback, the symbol A' designates the signal-voltage *amplification with feedback*. This symbol without the prime indicates signal-voltage amplification without feedback (A). The product $A\beta$ is occasionally used to indicate *feedback factor*. Thus,

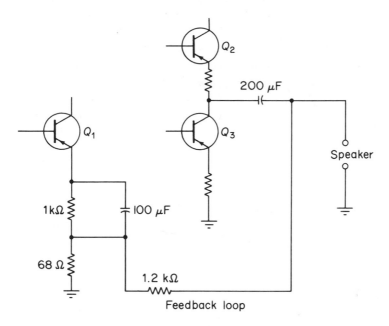

(a) inverse feedback, transistor
audio amplifier, capacitor coupled

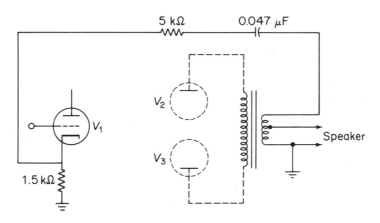

(b) inverse feedback, tube audio amplifier,
transformer coupled

figure 5-14 inverse signal-voltage feedback systems

$1 - A\beta$ is a measure of the feedback amplitude. In equation form, the signal voltage amplification with feedback is

$$A' = \frac{A}{1 - A\beta} \quad \text{or} \quad \frac{A}{1 + A\beta} \tag{5-12}$$

where A' = signal-voltage amplification with feedback
A = signal-voltage amplification without feedback
β = decimal equivalent of the percentage of output signal voltage fed back

When the feedback factor $-A\beta$ is much greater than 1, the signal-voltage gain is independent of A and, consequently, the formula for signal-voltage amplification with feedback becomes

$$A' = -\frac{1}{\beta} \tag{5-13}$$

Since distortion is reduced with inverse feedback, the amount of reduction can be indicated by an equation. Using D' to show distortion of output signal voltage with feedback, and D the distortion of the output signal without feedback, the equation is

$$D' = \frac{D}{1 - A\beta} \tag{5-14}$$

Another type of inverse feedback is *inverse-current feedback*, resulting from an unbypassed resistor in the emitter or cathode sections of transistors and tubes. In either case, signal-voltage changes across the resistor vary bias relationships between base and emitter or grid and cathode. The signal voltages across such a resistor act inversely to the increase or decrease of collector (or plate) current caused by the applied signal at the input, and degeneration results.

The negative feedback of the inverse-current type improves the frequency response of the amplifier stage, though the overall gain of the stage is reduced as with the voltage feedback system. A bypass capacitor upsets a flat frequency response because it has an increased reactance for lower-frequency signals and sets up signal-current variations across the resistor–capacitor network that act inversely and cause attenuation. For high-frequency signals, however, the reactance is low and the bypass effect is in force, maintaining a fairly steady-state direct current across the RC network. Where a wide-range frequency response is necessary in an amplifier, the unbypassed resistor is used for inverse-current feedback to assure equal processing for all signal frequencies.

5-7. video amplifiers

As shown in Fig. 5-1, video amplification follows the video detector in television receivers to bring the picture signals to the levels required for

application to the picture tube. Essentially, video amplifiers are audio amplifiers with special design features to extend the frequency-response range
over the span required to handle the video signals. Since the latter range
up to 4000 kHz, *peaking coils* are used to extend the frequency range, plus
direct coupling (or use of high-value coupling capacitors). In tube-type video
amplifiers, the plate load resistors are purposely decreased in value to help
improve frequency response. Tubes with low-value interelectrode capacitances are also used.

A typical video amplifier using a pentode vacuum tube is shown in Fig.
5-15. Resistor R_1 with capacitor C_1 form the bias network. Current flow

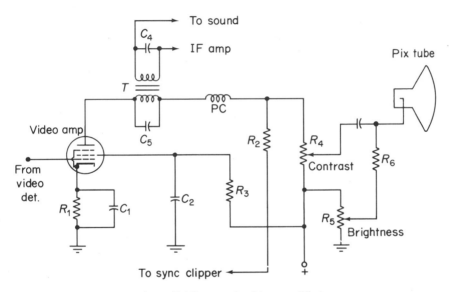

figure 5-15 pentode video amplifier

through R_1 sets up a voltage drop that is positive at the cathode, thus making
the grid negative in relation to the cathode. Resistor R_3 sets the proper screen-
grid voltage, and C_2 minimizes signal-voltage variations across the resistor.
The plate resistor R_4 has the amplified signal voltage dropping across it, with
this resistor also seving as a gain control for adjusting the contrast level
desired. Resistor R_2 taps the video signal for application to the vertical and
horizontal synchronization circuitry (see also Chapters 8 and 9).

The positive supply potential is tapped by R_5 and, in conjunction with
the series limiting resistor R_6, adjusts the bias for the picture tube. As the
cathode is made more positive, the grid becomes more negative and picture
brilliancy decreases.

Transformer T is for sound-take-off purposes. The primary (with C_5)
and the secondary (with C_4) form resonant circuits tuned to the sound IF
frequency of 4.5 MHz. The primary section also presents a high impedance

for the 4.5-MHz signal in the plate output section, thus minimizing the interference this signal would produce if it passed through the coupling circuitry to the picture tube.

The peaking coil (PC) is a small inductor which, because of its high reactance, isolates the shunt capacitances of the input and output circuits. It also permits use of a higher plate resistor for more gain. The higher gain, plus the reduced losses for higher-frequency signals because of the isolation of the shunt capacitances, compensates for the high series reactance to higher-frequency signals; since these are now amplified to the degree desired, they are considered to have been *peaked*.

Shunt peaking is also employed, as shown in Fig. 5-16. Here the peaking

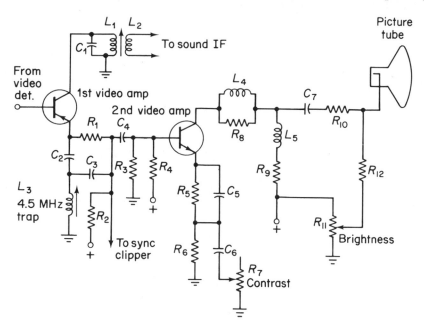

figure 5-16 transistorized video amplifier

coil L_5 shunts the input and output circuitry, and its inductance forms a parallel-resonant circuit in conjunction with the shunt capacitances. In consequence, a high impedance is formed tht effectively minimizes high-frequency signal loss by virtue of the low reactances formed by the shunt capacitances of the circuitry. The peaking coils are usually spaced well away from the chassis to minimize shunt-capacitance losses that might develop from too close a proximity.

The solid-state video system of Fig. 5-16 uses two transistors to realize the equivalent gain obtained by a pentode tube. The sound take off is in the collector circuit of the first video amplifier, and hence, does not attenuate the 4.5-MHz signal in the circuitry feeding the picture tube. Consequently, a 4.5-MHz trap is included in the emitter-follower circuit. Trap inductor L_3 and capacitor C_2 form a series-resonant circuit having a low impedance for the signals to which it is tuned and, hence, shunting these.

The series peaking coil L_4 is shunted by resistor R_8 to load down the inductance and thus lower the coil Q. Too high a Q, with consequent sharp resonance for a specific frequency, could produce transient oscillations that could cause interference on the tube screen. Such a high-Q coil could be pulsed into a ringing state by portions of the video signal.

The series peaking coil L_5 could also be placed below the load resistor R_9, since it would still present the required shunting effect and thus parallel the shunt capacitances.

questions and problems

5-1. Define the terms preamplifier, power amplifier, and modulator.

5-2. What type of signal losses are introduced by using resistance–capacitance coupling and transformer coupling?

5-3. A transformer is to match an impedance (Z_1) of 1875 Ω to another impedance (Z_2) of 5 Ω. What turns ratio must the transformer have?

5-4. For Problem 5-3, if the primary of the transformer has 75 turns, how many turns must the secondary winding have?

5-5. What design factors must be incorporated in a transformer to reduce losses of signal components?

5-6. Define *direct coupling* and explain why it is advantageous over *RC* or transformer coupling.

5-7. Explain the basic principles of treble and bass tone controls.

5-8. In an L pad an input impedance of 100 Ω is shunted by the pad resistor R_1 of 110 Ω. The output impedance is 500 Ω and the series resistor R_2 is unknown. Calculate the ohmic value of R_2.

5-9. In a T pad the input and output impedances are each 50 Ω. The 100-V input is to be attenuated for an output of 25 V. What is the value of R_1?

5-10. Describe a bridged pad and explain how the bridging resistor values are calculated.

5-11. Briefly explain what causes harmonic distortion.

5-12. Explain the differences between frequency distortion and phase distortion.

5-13. Describe at least four methods for minimizing distortion in audio amplifiers.

5-14. Explain two advantages of push–pull circuitry over the single-ended type.

5-15. Why does inverse feedback reduce distortion and circuit noises? Explain briefly.

5-16. In what manner does current-feedback circuitry differ from voltage feedback?

5-17. Explain what is meant by *feedback factor* and how this relates to amplification.

5-18. What is the purpose for using series and shunt peaking coils in video amplifiers?

5-19. Explain what measures are taken in a video amplifier to minimize the interference that could be caused by the 4.5-MHz sound IF signal appearing at the picture-tube input.

5-20. Briefly explain the electronic functions of the contrast control and the brilliancy control in video amplifiers.

5-21. Why are series peaking coils often shunted by resistors?

6

amplitude modulation and detection

6-1. introduction

The basic transmitting system was shown in block-diagram form in Fig. 2-1. A more comprehensive illustration of the complete system is shown in Fig. 6-1. As discussed, a crystal oscillator is used to establish frequency stability, though this circuit does not necessarily generate the fundamental carrier frequency. Often it generates a signal harmonically related to the final carrier frequency, but of lower frequency to permit use of a thicker crystal for greater efficiency and durability in temperature control. This applies in particular to high-frequency transmission practices, such as television or commercial shortwave services. If too thin a crystal is used, it would have to be operated at reduced ratings to minimize the creation of burned spots caused by arcing, or cracking. As shown in Fig. 6-1, frequency-multiplier Class C stages are used when necessary.

Additional buffer amplifiers are used to raise the signal power to the level necessary and, at the same time, to isolate subsequent stages from the initial crystal stage. This minimizes the possibility of load changes at the output stage affecting the stability of the oscillator circuit.

The RF power-output amplifier is also Class C for maximum efficiency. The modulator is an audio (or video) amplifier system, operated as Class A, AB_1, AB_2, or B, depending on power requirements, the amount of distortion that can be tolerated, and the particular design favored for the system in use.

Once a modulated carrier has been produced, it cannot be amplified

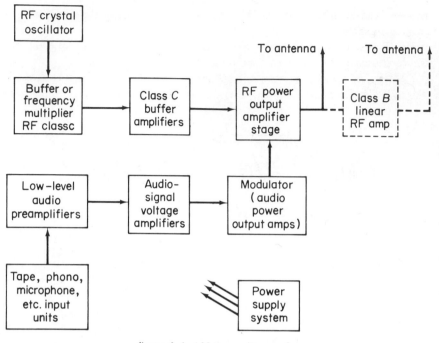

figure 6-1 AM transmitter system

additionally by successive Class C stages; instead, a Class B linear amplifier is used, as shown in Fig. 6-1. If the final Class C stage is modulated and no additional RF amplification is used for the modulated carrier, the system is referred to as *high-level modulation*. This term indicates that the modulation occurs at the *highest RF power level*, regardless of the type of modulation employed. When the modulated Class C amplifier is followed by one or more Class B linear amplifiers, the system is known as *low-level modulation*, since the modulation occurs at a level lower than the final output power from the last stage.

The various types of modulation and their appropriate circuitry are discussed in this chapter. In addition, the demodulation process is covered as it relates to the basic amplitude-modulation systems. Though the picture portion of television is also amplitude modulated, the complexity of the composite video signal necessitates circuitry in addition to that used for AM modulation and demodulation; hence, it is covered in detail in separate chapters devoted exclusively to the subject (Chapters 8 and 9).

6-2. solid-state amplitude modulation

Modern solid-state components can be used in low-powered, lightweight portable transmitter–receiver (*transceiver*) equipment, as well as in high-power fixed-location systems. The circuitry is essentially the same and the basic principles relating to low-power equipment also hold for the high-power units. Transistors (as well as tubes) can be paralleled or used in push–pull for increased power over that obtainable from a given unit.

A typical amplitude-modulated Class C *pnp*-transistor stage is shown in Fig. 6-2. The constant-amplitude carrier is applied to the base input system

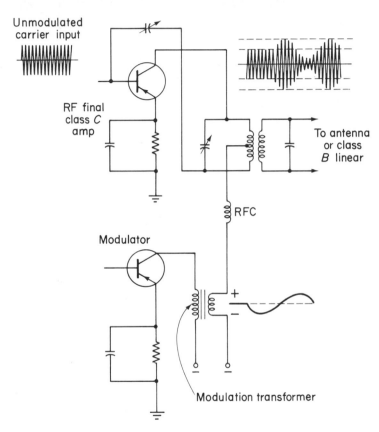

figure 6-2 single-ended transistor amplitude modulation

of the Class C final and is amplified in the collector circuit. Note, however, that the required reverse bias for the collector (negative polarity) is applied in series with the secondary of the modulation transformer. When the modulator produces an amplified output signal, a varying voltage drop occurs across the secondary, representative of the ac-type signal output. This change of voltage across the secondary of the modulation transformer *adds* and *subtracts* from the voltage applied to the collector, thus causing the amplitude of the carrier to change in accordance with the voltage changes across the modulation-transformer secondary. Since it is instrumental in causing an amplitude change, this type of modulation is known as *amplitude modulation,* or AM.

When an output alternation from the modulator is negative, it adds its voltage to that of the collector's negative-potential reverse bias and, hence, the carrier amplitude increases, as shown in Fig. 6-2. For a positive output alternation, the voltage across the modulation-transformer secondary opposes that of the applied reverse bias; hence, the amplitude of the carrier signal declines. These amplitude changes represent the carrier signal plus the sidebands generated by the modulation process, as more fully described in Chapter 2.

With a Class C amplifier having a fixed load impedance, the circuit resistance remains constant, and the power of the carrier varies as the square of the voltage. Thus, during 100 per cent modulation, the peak output power of the carrier wave from a Class C amplifier attains a value four times that of the unmodulated carrier power. Hence, for 100 per cent modulation, the amplitude of the carrier signal varies between zero and twice the amplitude of the unmodulated carrier.

During modulation, the *average* collector current (or plate current in a tube circuit) does not change, because successive increases in current caused by the modulation process are balanced by identical current decreases. For 100 per cent modulation, the output power of the *modulator* must be one half the Class C input power. *Input power* refers to the product of the Class C amplifier dc voltage and dc current. (The *signal input* to the base of the Class C amplifier transistor is termed *excitation,* and is unrelated to the dc potentials and currents.)

When voice, music, or video signals are broadcast, the percentage of modulation varies constantly because of the amplitude changes that occur for different volume levels of the modulating signals. The modulation percentage is defined in relation to how much less the modulating power is than one half the carrier amplifier input power. In adjustment or test purposes, a constant-amplitude modulating signal is used, and the following formula applies:

$$P_a = \frac{m^2 P_i}{2} \tag{6-1}$$

where P_a = required modulator audio or video power
$\quad\quad m$ = percentage of modulation in decimal form, as 0.5 for 50 per cent modulation
$\quad\quad P_i$ = Class C input power (dc)

Thus, if the input power to the Class C stage is 1000 W, for 100 per cent modulation we obtain the following audio-power designation:

$$P_a = \frac{1 \times 1000}{2} = 500 \text{ W modulating power required}$$

Similarly, if the modulation is 0.25 and the applied voltage to the Class C is 1600 V at 0.5 A, we obtain the following for P_a:

$$P_a = \frac{0.25 \times (1600 \times 0.5)}{2} = \frac{200}{2} = 100 \text{ W}$$

For a given set of conditions, we solve for the percentage of modulation. [See Eq. (2-1).]

The amplitude variations of the carrier shown in Fig. 6-2 represent power changes in the composite carrier signal, which include the sideband components, as described in Chapter 2. During amplitude modulation, the carrier amplitude does not vary, but the power of the sideband signals will alter proportionately to the amplitude levels of the modulating signal. In collector (or tube plate) modulating processes, the sideband signal power *is supplied by the modulator.*

The field-effect-transistor modulating system shown in Fig. 6-3 compares to the transistor type of Fig. 6-2. Instead of the collector voltage variations, the

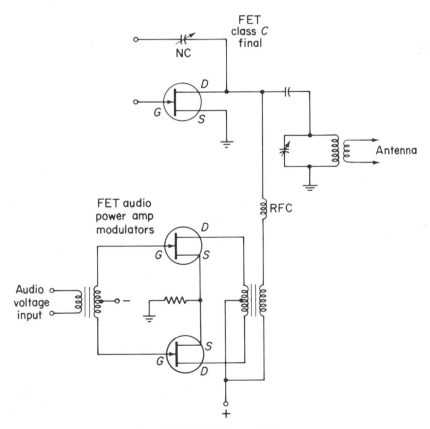

figure 6-3 FET drain modulation

drain voltage is altered by the modulating signal developed by two FET units in push–pull, as shown. As the modulating signal appears across the secondary of the transformer, it alters the applied drain voltage and, in consequence, changes the carrier amplitude accordingly.

As with the transistor unit of Fig. 6-3, the turns ratio of the modulating transformer is determined by using Eq. (5-1) to match the impedance of the modulator output to that of the Class C amplifier. The modulated carrier appears across the *tank circuit* (the term usually applied to the output resonant circuit of the Class C amplifier) and is transferred to the secondary for application either to an antenna system (high level) or to a Class B linear amplifier for low-level modulation (see Section 6-1).

The Class C stage could, of course, also be operated in push–pull and modulated as with the single-ended stages. A typical circuit using push–pull for both the Class C amplifier and the modulator is shown in Fig. 6-4. In

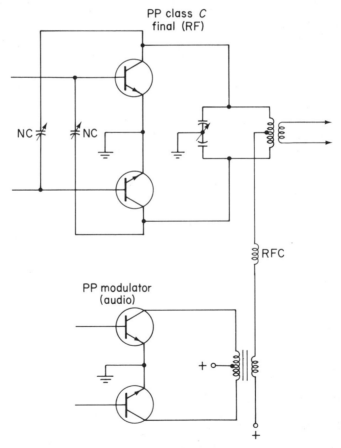

figure 6-4 transistor push–pull modulator and class C

contrast to Fig. 6-2, *npn* transistors are used, though *pnp* types could be substituted with a change in the forward- and reverse-bias potentials.

The radio-frequency choke (RFC) prevents leakage of the RF energy to the modulator or power-supply system. It presents a high reactance for RF but a low reactance for the lower-frequency modulating signals. The tank circuit uses split-stator capacitors, with the rotor sections at ground potential for safety during tuning. Cross-neutralization circuitry is used, as was shown for tube types in Fig. 4-14.

Push–pull increases power output and reduces distortion by providing for balanced circuitry. In Class C the system decreases undesired RF harmonics, while for the modulator it reduces the generation of undesired harmonic components, as discussed in Chapters 4 and 5.

6-3. tube-type AM

As with transistors, tubes are found in Class C applications, ranging from low power to high power (with forced-air cooling or water circulation around the anode for cooling purposes). Operational principles for the modulators and Class C amplifiers follow closely the explanations given for the solid-state types. A typical example of push–pull tube circuitry is shown in Fig. 6-5.

Even though this is still amplitude modulation, the term *plate modulation* is often used to distinguish this system from that of *grid modulation* discussed next. (Similarly, the transistor type could be termed *collector modulation* when AM is developed by applying the modulating signal to the collector circuitry, or *drain modulation* when AM is developed in this FET circuitry.)

For the circuitry shown in Fig. 6-5, conventional split-stator tuning is used, with appropriate cross neutralization to prevent the generation of spurious harmonic components. Again, a modulating transformer links the output from the modulator to the dc input of the Class C amplifier. A current meter (I) is usually placed in series with the plate circuitry, as shown, so that currents can be read during tuning processes and a constant check provided on normal operation. The shunting capacitor bypasses RF and avoids meter damage.

As with other Class C amplifiers, tuning for resonance is aided by watching for a dip in plate current as indicated by the ammeter. Often initial tuning is performed without the antenna system or other load applied. After the circuit has been tuned to resonance, the load circuit is applied and coupled for maximum current reading on the meter. The load circuit is then tuned to resonance as indicated by a slight decrease in the meter reading.

When the tank circuit is tuned off resonance, plate (or collector) currents reach maximum values. When off resonance, the impedance of the parallel-resonant circuit declines and, in consequence, power supply currents increase. With high-power Class C amplifiers the currents can reach excessive propor-

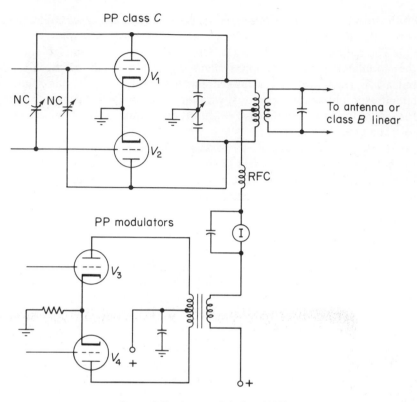

figure 6-5 plate modulation (AM)

tions when the circuits are tuned off resonance; hence, some protective measures are usually incorporated to limit plate-current flow during the tuning process. Switches decrease the applied voltage to safe values while adjustments are being made. With a load, the Q of the tank circuit is extremely high and excessive voltages build up across the tank circuit and can cause arcing in the tuning capacitor or across the tuning inductors. To minimize arcing, capacitor plates and coil turns are usually spaced apart to the degree required by the particular voltages utilized. Under normal load conditions when the antenna system or the subsequent amplifier stage is attached, the Q drops to a value as low as 10 to 15.

Neutralization procedures are the same as described in Section 4-9 for Class C amplifiers. Since two neutralizing capacitors are used in push–pull circuitry, each must be adjusted alternately until the desired neutralization has been accomplished.

As shown in Fig. 6-6, it is also possible to modulate a Class C amplifier by applying the modulating signals to the input. For Fig. 6-6 this involves *grid modulation*, and if pentodes were used for the Class C amplifier tubes, screen-grid modulation could also be employed.

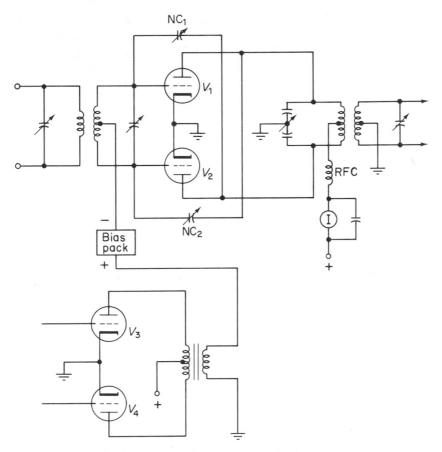

figure 6-6 grid-modulation circuitry

When modulating other grids, the circuit factors are the same as those described for grid modulation. The modulator used for grid modulation need only have a fraction of the power output required for plate modulation. In grid modulation, the secondary of the modulation transformer adds to or subtracts from the bias voltage supplied by the bias pack shown in Fig. 6-6. Since a bias change is instrumental in causing a plate current change, any signal variations in the modulation transformer cause amplitude changes in the carrier signal of the anode circuit. Thus, if a positive alternation appears across the secondary of the modulating transformer, it would subtract from the bias supplied by the bias-pack power supply and, hence, plate currents would rise, increasing amplification and signal amplitudes. For a negative voltage across the transformer secondary, the bias would increase and less current would flow through the tubes, decreasing signal amplitude.

For grid modulation, the RF carrier signal applied to the input section of the Class C amplifier must have an amplitude such that it extends approximately between the limits shown in Fig. 6-7 at the bias indicated. This excitation placement on the characteristic curve is critical to ensure that the modulation process can swing the carrier signal both above and below the unmodulated value without driving plate currents below zero or into the saturation region.

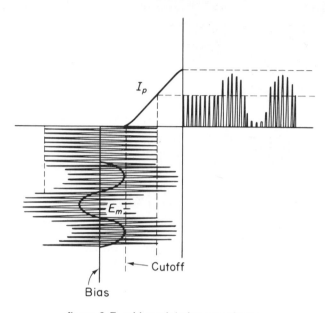

figure 6-7 grid-modulation waveforms

Plate current, in pulses as shown, produces the modulated carrier in the output resonant circuit because of the flywheel effect produced therein. The interchange of charged energy between the capacitance and inductance of the resonant circuit causes positive and negative signal alternations, thus completing the composite amplitude-modulated waveform.

Because the excitation to the Class C amplifier is reduced for grid modulation, the power output and efficiency of the output stages are below what they would be for plate modulation. In plate modulation the sideband power was furnished by the modulator, but in grid modulation the Class C amplifier system must furnish the sideband power. Thus, the grid-modulation method has some disadvantages with respect to plate modulation. An advantage, however, is the low modulating power needed, which results in considerable saving in high-powered modulators.

Note in Fig. 6-7 that the modulating signal E_m shifts the bias line in accordance with the waveshape of the modulating signal across the transformer secondary. Hence, as the bias line shifts, it takes the carrier amplitude with it, but does not alter the average amplitude of the carrier at the input section of the Class C amplifier. The amplitude of the modulating signal is, therefore, critical and must not be so excessive that it causes the carrier to shift below the cutoff point or into the positive-grid region of the tube.

With triode tubes there is greater latitude for carrier shift than there would be in the pentode tube. In the triode, the characteristic curve is not as steep as that of the pentode. Even with triode operation, however, the danger of overmodulation is greater with this modulation process than it is with plate modulation.

6-4. AM detection

The unidirectional current characteristics of diode (whether germanium, silicon, or vacuum tube) make it ideal in communication electronics for extracting from an amplitude-modulated carrier the information transmitted. Thus, the diode detector (demodulator) is widely used for audio and video detection, as well as for obtaining automatic volume- or gain-control voltages, as covered later.

A typical AM detector circuit is shown in Fig. 6-8 and represents the demodulation system following the last IF amplifier (see Fig. 4-1). The resonant-circuit transformer is tuned to the IF signal; hence, the secondary obtains the RF signal information as shown. For positive signal alternations across the secondary, electron flow is in the direction shown by the arrows, producing a voltage drop across R_1. During negative alternations across the secondary, the diode does not conduct; hence, no current flows in the circuit. Thus, the diode rectifies the RF signal energy, just as low-frequency power supplies utilize diode rectification. The result, as shown in Fig. 6-8, is that the amplitude variations of the pulsating direct current are converted to an average direct current having an amplitude that changes according to the audio- or video-frequency amplitude used to modulate the carrier.

As with conventional power supplies, a filter network is employed in the demodulator to obtain direct current from the pulsating signals produced during rectification. For Fig. 6-8 the filtering network consists of C_3 and R_1, with the latter also acting as the volume control. The filtered voltages across R_1 are unidirectional, but the variations are the audio or video signals. Capacitor C_4 converts these to an ac-type signal, with positive and negative alternations.

For an unmodulated carrier (time t_1) the IF signals arriving a the detector have a constant amplitude; hence, the average value of the pulsating direct current appearing across R_1 also has a constant amplitude. For any positive alternation that increases above the level representing zero modulation, a larger charge appears across the filter capacitor C_3 and the dc voltage am-

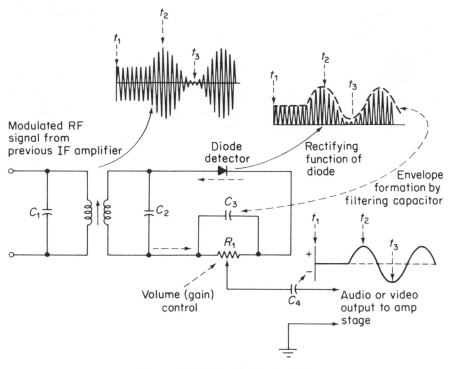

figure 6-8 diode detector for AM

plitude across R_1 rises. Thus, at the positive peak of the modulated signal t_2, the voltage across R_1 reaches a maximum, representing the positive peak of an audio alternation at the output. As the modulated RF signal declines to a value below that of the unmodulated state, the voltage drop across R_1 declines, producing the negative alternation of the audio output, as shown.

The time constant (RC) of the filter network is sufficiently long to prevent capacitor discharge between rectified alternations, yet short enough to permit the voltage across R_1 to follow that of the modulated signal. Capacitor C_3 thus has a high reactance for the audio signal and permits its development across R_1. For the RF signal, however, C_3 has a low reactance and, hence, offers a bypass effect.

The demodulator of Fig. 6-8 could operate as an individual receiver (for earphone operation) if an antenna lead were connected to the secondary winding. Without preliminary amplification, however, reception is relatively weak and selectivity is poor. Local stations, however, are received with good earphone volume.

Diode detectors have a low impedance and load down the last IF stage sufficiently to decrease selectivity to some extent. Hence, C_1 and C_2 tune over a wider selectivity range than other IF stages in a given receiver.

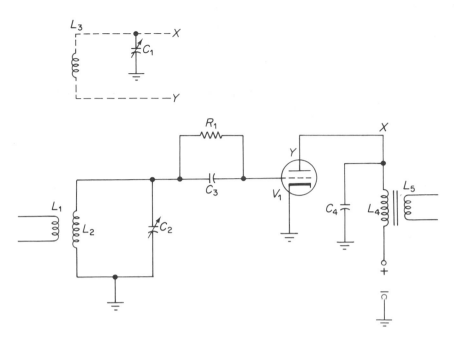

figure 6-9 triode tube detector

Though the solid-state diodes are widely used in detection systems, the tube type is shown in Fig. 6-9 for reference and comparison purposes. A transistor could be substituted in this circuit and the operational characteristics about to be described would still apply. For transistors, FET's, and other solid-state units, appropriate bias voltages would, of course, have to be applied, plus impedance-matching transformers. A pentode tube could also be used instead of the triode shown. For either the tube triode or pentode, the detection system is known as a *plate detector*. Again, the signal (modulated RF) is obtained from an antenna system or a previous amplifier.

The modulated signal appears between the grid and cathode of the tube, and for positive alternations grid current flows. The unidirectional current develops a filtered voltage drop across R_1, as was the case for the solid-state diode discussed. The voltage for R_1 is negative toward the grid, thus establishing the grid bias for the tube. Capacitor C_3 and resistor R_1 are selected to produce a bias at the approximate cutoff point of the tube (approximately 0.00025 μF, and 1 megohm (MΩ) for the broadcast band).

The modulated waveform between cathode and grid also affects plate current flow, with all positive alternations of the input signal causing plate current to flow in pulses. This process, in effect, again is a rectification principle, since the plate-current pulses are unidirectional. Capacitor C_4 in the anode circuit has a bypass effect on the RF signal energy, and will also charge to the peak values of the plate-current pulses. Therefore, this capacitor forms the average value of the plate-current pulses and, thus, procures the audio component from the rectifying process.

The plate detector operates on the rectilinear portion of a characteristic curve; thus, the ratio between the RF input signal voltage and the audio-frequency signal output voltage is virtually linear. Hence, this demodulator is often termed a *linear plate-circuit detector*. If the signal input has a very low amplitude, some distortion results because of operation on the curved or nonlinear portion of the characteristic curve of the tube. This drawback is eliminated by the diode-type detector. The linear plate detector has good sensitivity, however, and some amplification of the signal occurs because of the grid and anode sections of the triode.

Selectivity is not lowered, as with the diode detector, since the plate-circuit detector has a high input impedance, thus minimizing the loading effect on a tuned resonant circuit at the input. While the tube-type detector has some specific applications in special commercial or experimental fields, the diode detector is by far the most popular because of its simplicity and its ability to handle higher input signal potentials.

For Fig. 6-9, the *regenerative detector* can be formed by including a feedback loop, shown in dotted form (L_3 and C_1). The feedback circuit is connected to the plate lead points marked y and x, with the lead between these points opened when the feedback circuit is connected. The inductor L_3 feeds back a portion of the signal energy in the anode circuit to the secondary of the input transformer L_2.

When the fed-back signal polarity coincides with that at the secondary, signal reinforcement occurs and circuit efficiency is raised. Selectivity and signal sensitivity also increase because of the reduction of circuit resistance. This feedback system is termed *regeneration*, and the degree of regeneration can be altered by changing the spacing between L_2 and L_3, or varying the capacitor C_1 (which replaces C_4). A variable resistor could also be used in the voltage lead to the anode, since a change of anode voltage affects the degree of signal amplification in the anode and, hence, the amplitude of the signal fed back.

If the regenerative (positive) feedback is gradually increased, a point will be reached where the amplitude of the fed-back signal is so great that the circuit breaks into self-oscillation and, hence, generates its own signal. When this occurs, demodulation of AM signals without severe distortion is virtually impossible. This is one of the disadvantages of the regenerative detector—its critical adjustment for maximum performance without breaking into oscillation. When maximum regeneration without oscillation has been established, the inherent instability of the system soon causes the detector to break into oscillations, requiring readjustment. Hence, the regenerative AM detector is not used in modern circuitry, except in experimental types.

One application for the regenerative detector, however, is its use in the oscillating state as a detector for transmitted code signals having no amplitude modulation characteristics, as described next.

6-5. CW and ICW coding

In the transmission and reception of the Morse code in shortwave commercial and governmental applications, two methods are employed. One type uses a carrier of *fixed amplitude and frequency*, but interrupts its transmission to form short and long sections representing dots and dashes, as shown in Fig. 6-10. This type of transmission is known as CW, for *continuous waves*,

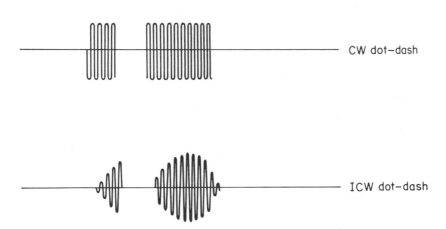

figure 6-10 CW and ICW signals

to indicate that the amplitude is continuous and does not vary. Another method is to use a modulated carrier and to interrupt it, as also shown in Fig. 6-10. This type is known as ICW (*interrupted continuous wave*), indicating the continuous amplitude is not maintained but rather is varied (interrupted from its constant amplitude). This ICW signal utilizes a broader band than CW because the modulating process generates sidebands, while the unmodulated CW is a narrow-band-type transmission.

For the ICW transmission, any AM radio is capable of demodulation. Thus, if the ICW is modulated by a 400-Hz tone, the audio from the radio consists of short- and long-duration 400-Hz tones representative of the Morse code. For the CW transmission, however, an ordinary radio is not suitable for reception since there is no modulation to detect. Instead, detection consists of generating an independent signal and applying it to the detection circuitry in conjunction with the transmitter CW signal. The two are made to *heterodyne*, that is, beat together to produce a new signal having a frequency that represents the difference in frequencies between the CW signal and that generated by the detection process.

In Section 5-4 it was pointed out that circuitry having nonlinear characteristics produces harmonic distortion consisting of undesired signals not found in the original. In Chapter 2 the generation of new signals (sidebands) by modulation was also discussed. Thus, if two signals are injected into a circuit that has a high degree of nonlinear characteristics, the output signals consist of the intial two signals, plus additional signals having frequencies equal to the sum and difference frequencies of the original two signals. The additional signals are produced by the heterodyning, or beating together, of the original two signals. Thus, if one of the original signals has a frequency of 2000 kHz, and the other signal has a frequency of 1999.5 kHz, the heterodyning of the two generates an aditional signal having a *difference* frequency of 500 Hz (0.5 kHz). At the circuit output, however, the original two signals would also appear, as well as the sum frequency of the two (3999.5 kHz). The undesired signals are filtered out by appropriate circuitry.

The heterodyning process is not used solely for the detection of CW signals, since its principle is also utilized in almost all modern AM radio, FM, television, shortwave, and other receivers using the superheterodyne principle (see Fig. 3-8 and the related discussion in Sections 3-4 and 6-7). When the local oscillator of such a receiver mixes its signal with that of the incoming carrier signal, sum and difference signals are again produced, since a heterodyning process is involved. In superheterodyne receivers, however, the intermediate frequency (IF) signal finds resonance in the IF amplifiers and is thus accepted by the system. The resonant circuits, however, reject the original signals and the sum signals. Since the oscillator is made to tune in synchronization with the mixer stage (and RF stage, if one is used), the same difference (IF) signal is obtained (usually 455 kHz for AM radios).

For detection of a CW signal, the regenerative-circuit detector of Fig. 6-9 (or its solid-state equivalent) is used. The detector circuit is adjusted for the oscillatory mode to generate a signal to mix with the incoming CW carrier. If the oscillator signal's frequency is near that of the CW frequency, an audio tone is generated. If, for instance, the CW signal is 30,000 kHz and the detector's oscillation signal is 30,001 kHz, the difference frequency of the signal produced is 1 kHz (1000 cycles), which is an audible tone.

Because of the frequency difference between the incoming carrier signal and that of the oscillating detector, point-to-point addition of the two involves phase changes in the signals. Consequently, the resultant signal will have greater amplitudes when the phase is close and decreasing amplitudes as a greater phase difference is encountered. This constitutes an amplitude modulation of the difference-frequency signal and, hence, is detected as such to produce an audible output from the regenerative detector. Since amplification also occurs, the output signal is much greater in amplitude than that obtained from the diode detector.

As the detector's signal frequency and that of the incoming signal are separated more and more, an increasingly higher audio-signal frequency is produced. As the difference frequency rises to values of 15 kHz or higher, they soon become inaudible to the average ear. If the detector's signal frequency is altered to bring it closer to that of the incoming signal, an increasingly lower audible signal is produced. When the two signals coincide, a null or zero output is reached.

The oscillating-type detector is also called an *autodyne* detector. If CW signals are to be received by the diode AM detectors, a separate oscillator would have to be used to provide for a signal for injection into the detector circuit for heterodyning with the incoming constant-amplitude CW signal.

Since the autodyne detector oscillates, it is capable of radiating a signal such as to cause interference with other communication services. Hence, the circuit should be well shielded and operated with an isolating RF amplifier stage between it and the antenna system.

The regenerative detector can be converted into a *superregenerative* type by injecting a supersonic frequency signal in series with the regeneration inductor. Such a supersonic signal is obtained from an additional oscillator generating a signal of 25 or 30 kHz. The anode supply potential is set slightly below the point where oscillations normally occur. Thus, the supersonic signal will alternately add positive and negative potentials to the supply voltage. The positive alternations increase the anode voltage and cause oscillations, while the negative alternations decrease anode potentials and stop oscillations. Thus, such a demodulator is thrown in and out of oscillation at the supersonic rate. Since this rate is above the audible range it cannot be heard at the output. It does, however, produce an oscillatory state during one half the operating time and, hence, realizes the full advantages of such a system in terms of higher efficiency and signal sensitivity. Since the oscillating state is periodic and not sustained, no heterodyning occurs and, hence, no beat frequency (difference-frequency signal) is produced. Thus, for voice or ICW reception, the superregenerative detector offers some advantages over the other types.

When the superregenerative detector is not tuned to a station, the absence of signal input causes a loud hiss. This noise is the result of the rapidly changing supply potential at the supersonic rate and its effect on current flow. During reception of a signal, this noise is suppressed. The superregenerative receiver has poor *selectivity*, despite its high efficiency and good signal sensitivity. Its principal applications are for ICW reception or for reception of AM signals that are transmitted in a sector of the frequency spectrum uncrowded by other broadcasting facilities.

6-6.   *automatic volume control*

For maintaing a volume level in radio reception as preset by the listener, a system known as *automatic volume control* (AVC) is used. Another system, *automatic gain control* (AGC) is similar in operational function, but is used for maintaining the level of the picture signal as preset by the television viewer. The AGC circuitry is covered in Chapter 8. Both systems prevent strong signals (local stations) from overloading receivers and aid in reducing fluctuations of reception that might arise from local interference.

In the AVC system a voltage is obtained from the AM detector circuit and connected to the RF–IF input sections to regulate the gain of these amplifier stages. Thus, if the radio listener has set the volume control to his desired level and the signal fades slightly, the AVC signal automatically increases the gain of prior stages and thus compensates for the signal decline. If the incoming signal increases, more bias voltage is obtained from the AVC circuit and this decreases the gain of the earlier amplifier stages, again compensating for the undesired change in volume level.

The circuitry for obtaining the AVC voltage from the diode AM detector is shown in Fig. 6-11. The direction of current flow through the detector diode

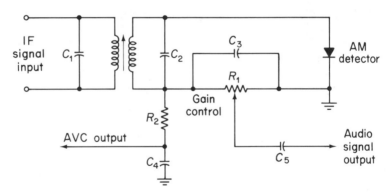

figure 6-11 AVC and diode detector

and resistor R_1 causes a voltage drop across R_1 that is negative toward the bottom of the transformer secondary and positive toward the diode. The incoming AM carrier will create an average value of voltage across this resistor, though the voltage will vary in accordance with the amplitude changes of the incoming signal. Thus, when a lead is connected to the negative side of the resistor, a voltage is obtained that represents the average value produced by the detection process. Resistor R_2 helps isolate the detector system from the stages to which the AVC voltage is applied, and in conjunction with capacitor C_4 forms a filter network for elimination of the audio-frequency signal components that would cause the average dc value to fluctuate. Capacitor C_4 has a low reactance for the audio voltage for bypass effect, and, hence, cannot be connected directly to the negative side of R_1, because it would shunt the audio signals to ground and kill the audio output at C_5.

The AVC output voltage is constant in amplitude and representative of the average value of the received carrier. In tube RF–IF stages it is connected to the grid circuits for changing bias and thus tube gain. For transistor stages, it alters the forward bias at the input (see Fig. 4-6).

6-7. mixer-converter

As mentioned in Section 3-4 and shown in Fig. 3-8, the incoming modulated RF signal is applied either to an RF-amplifier stage or directly to a *mixer* or *converter* circuit, which mixes (heterodynes) the incoming carrier signal with that generated in a local oscillator. The result is the production of an IF signal that remains the same regardless of the frequency of the station to which the receiver is tuned. This permits selection of the optimum degree of selectivity within the IF stages for obtaining smooth tuning characteristics with proper bandwidth for all stations. This system is known as the *superheterodyne* principle (see Section 6-5) and is utilized in most communications receivers. Performance if far superior to that which would be obtained for a series of RF stages in which each must be tuned to the frequency of the incoming signal, with resultant variations in Q.

The mixer-converter circuit that produces the IF signal has detection characteristics. Hence, it is sometimes referred to as the first detector of a superheterodyne receiver, whereas the diode demodulator following the IF stages is termed the second detector.

A typical transistorized mixer stage is shown in Fig. 6-12. This circuit uses the *autodyne* (self-oscillating) principle wherein the mixing process, as well as the signal-generating factor, is accomplished by the single transistor circuit. If the RF signal input is obtained from an RF amplifier stage, it would be coupled to the mixer by the transformer arrangement between L_1 and L_2. If no RF stage is present, L_2 is wound over a *loopstick* (a ferrite rod), and the assembly acts as the antenna. In such an instance, L_1 can be attached to an outdoor antenna for increased signal pickup.

Capacitor C_1 is the variable-tuning capacitor and is ganged to the variable capacitor that tunes the oscillator (C_5). The broken-line interconnections between C_1 and C_5 indicate that the capacitor rotor sections are ganged together for rotation by a common shaft. Capacitors C_2 and C_6 act as *trimmers* for tuning out resonance difference between the two circuits.

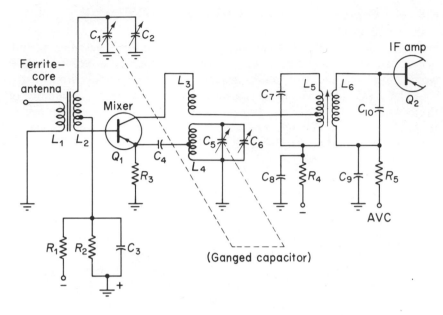

figure 6-12 mixer (converter) circuit

A feedback coil L_3 couples some of the amplified signal in the collector of Q_1 to the oscillator resonant-circuit inductance L_4. The coupling is polarized to produce positive (regenerative) feedback for oscillations. The RF signal in the oscillator resonant circuit L_4 and C_5 is coupled back to the emitter circuit via C_4.

The RF signal appearing between the base and emitter of Q_1 heterodynes with the oscillator signal because of the nonlinear operational characteristics established by the applied bias operating the transistor near cutoff. After the mixing process the sum and difference signals present at the output of Q_1 are rejected by the resonant circuits feeding the IF amplifier, since these are tuned to the IF-signal frequency only.

The inductor L_5 is tapped to obtain the proper impedance match between the low impedance of the collector circuit of Q_1 and the relatively high impedance of the parallel-resonant circuit composed of C_7 and L_5. Similarly, L_2 and L_4 are tapped for a reduction of impedance to meet that found in transistor circuitry.

An IF frequency of 455 kHz has become virtually standard for the AM broadcast band. Thus, the oscillator in the converter circuit differs from the incoming RF-signal frequency by this 455 kHz. Hence, if the frequency of the station being received is 800 kHz, the oscillator would generate a 1255-kHz signal to produce the proper IF signal. (The frequency of the oscillator is tuned *above* the frequency of the incoming RF signal to improve the tuning tange of the oscillator over that which would prevail if the oscillator's frequency were 455 kHz below that of the incoming signal frequency. Thus, for a 1000-kHz station, the oscillator's signal becomes 1455 kHz.)

questions and problems

6-1. Explain what is meant by a *buffer* amplifier in the Class C amplifier stages of a transmitter.

6-2. Explain the difference between low-level modulation and high-level modulation.

6-3. Briefly describe the electronic function of a modulating transformer.

6-4. Define the term *input power* to a Class C amplifier.

6-5. The input power to a Class C plate-modulated stage is 50,000 W. For 100 per cent modulation, what must be the audio-output power of the modulator?

6-6. For test purposes, the modulation of a Class C amplifier is held at 75 per cent during AM collector modulation. If the carrier power is 25,000 W, what must the audio modulator produce?

6-7. When modulation reaches 90 per cent and the applied voltage to the Class C is 10,000 V, with 4 A of current flowing, what audio power must the modulator deliver at this time?

6-8. In collector (or plate) modulation, where is the sideband power obtained? Explain briefly.

6-9. Briefly explain the tuning procedures for the Class C section of an AM transmitter.

6-10. Briefly explain why excitation and modulation factors are more critical for transistor base or tube grid modulation.

6-11. In grid modulation, where is the sideband power obtained? Explain briefly.

6-12. Explain briefly how the rectifying principle is employed by a diode during the demodulation process.

6-13. What type of detection systems are necessary for reception of CW and ICW signals?

6-14. Illustrate, by schematic drawings, two methods for controlling regeneration in a detector utilizing positive feedback.

6-15. Define the term *autodyne detector*.

6-16. Explain briefly the functional characteristics of the superregenerative detector system.

6-17. What disadvantages, if any, are found in the superregenerative detector as compared to other types?

6-18. Briefly explain the advantages of using AVC and AGC circuits in receivers.

6-19. How is the bias voltage (obtained from an AVC system) kept from fluctuating in accordance with the amplitude modulation on the carrier?

6-20. To what circuits of what stages are AVC voltages applied?

6-21. Describe the circuit system making up the superheterodyne receiver.

6-22. In what manner are the sum-frequency signals from a mixer minimized in the IF stages?

6-23. In a superheterodyne receiver the RF stage is tuned to 650 kHz. With an IF of 455 kHz, what is the mixer-oscillator signal frequency?

7

frequency modulation and detection

7-1. introduction

The processes for frequency modulation and detection are much more complex than amplitude modulation because of the variety of modulation systems, the types of services furnished by broadcasters, and the much higher frequency spectrum involved. In amplitude modulation the problem of frequency stability of the carrier presents no serious problems, since carrier frequency is not altered during modulation. For FM, however, the necessity for changing carrier frequency with modulation requires a variable-frequency oscillator, and frequency-control problems are increased.

For frequency modulation, any one of several systems may be used for the direct or indirect process, as more fully described later. In addition, a broadcast station may transmit one or more of the services shown in Fig. 7-1. In standard FM (88 to 108 MHz) a station may only be using monaural broadcasting. Another station may be broadcasting stereo multiplex (two separate channels, left and right) combined into a single transmission. Some mono- or stereo-broadcasting stations may also be involved with SCA (*Subsidiary Communications Authorization*) multiplex transmissions, where the main channel is used for general-public broadcasting, and the sub-channel of SCA for background music or other private services to paying subscribers.

In addition, a form of narrow-band frequency modulation is used by some commercial services. Also, the sound accompanying standard television

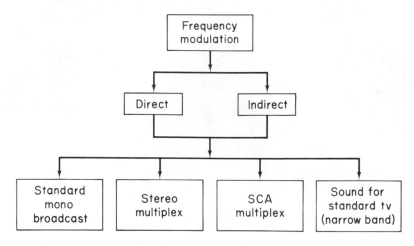

figure 7-1 FM systems

broadcasts is frequency modulated, though again with much less bandspread than used in the 88- to 108-MHz broadcasting.

To facilitate clarification of the variety of systems and methods found in FM, the initial sections of this chapter are devoted to an analysis of the complete systems, after which individual circuitry is analyzed. Reference should also be made again to Chapter 2 for the discussions on angle modulation, as well as those on frequency- and phase-modulation processes.

7-2. direct frequency modulation

In *direct* frequency modulation, as the term implies, the RF carrier is modulated directly by the audio component, in contrast to the *indirect* method (discussed later) wherein the variable-frequency oscillator is modulated by an AM-to-FM conversion process. In the direct system several processes are available, and one method is illustrated in Fig. 7-2.

As shown, the carrier signal is generated by a variable-frequency oscillator, which could be of the Hartley type discussed in Chapter 3. This oscillator is tuned between 4.5 and 6 MHz for transmission within the standard FM band of 88 to 108 MHz, using the frequency multipliers shown. Thus, if final transmission is, for instance, to be 90 MHz, the variable-frequency oscillator would be tuned to 5 MHz, the first tripler would produce 15 MHz, the second tripler 45 MHz, and the doubler brings this to the final frequency of 90 MHz. Thus, in any particular system the number and degree of frequency multipliers depend on the initial oscillator frequency and the final carrier frequency.

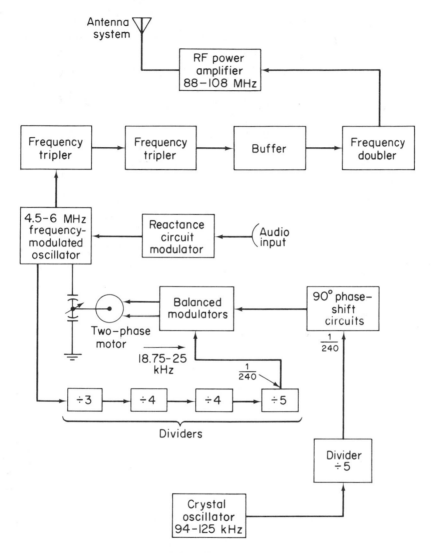

figure 7-2 direct frequency modulation

A reactance circuit is used to vary the frequency of the variable oscillator to produce frequency modulation. The reactance circuit incorporates a tube, transistor, or other solid-state device that has varying inductive or capacitive characteristics related to the applied voltage, as discussed more fully later. Since a variable-frequency oscillator generates the carrier (a necessity for frequency modulation of this type), some means must be utilized to stabilize the carrier frequency so that it will not drift from its assigned place in the spectrum. For the transmitter shown in Fig. 7-2 stability is achieved by use of a motor-control system that compares the local-oscillator frequency with that of a crystal oscillator. If the variable-frequency oscillator drifts away from its normal frequency, a correction voltage is obtained from a balanced-modulator circuitry, which, in turn, is applied to a two-phase motor. The motor then turns and rotates the variable tuning capacitors of the oscillator and brings the oscillator back to the proper center frequency.

Since the variable-frequency carrier oscillator is modulated, the carrier swing each side of center is minimized before application to the motor-control section. This is done by a series of dividers that bring the oscillator's signal frequency to $\frac{1}{240}$ of its original value. During such division, the *extent* of frequency deviation also decreases proportionately; hence, the signal appearing at the balanced modulators is virtually constant in frequency, since its deviation is now negligible.

As shown in Fig. 7-2, four dividers can be used to bring the final signal to a frequency between 18.75 to 25 kHz. The crystal oscillator used to stabilize the carrier frequency generates a signal having a frequency between 94 and 125 kHz. This signal is divided by 5 so that its frequency matches that obtained by dividing the variable-frequency oscillator. The signal from the crystal-oscillator source, however, is shifted in phase by 90° with respect to the signal from the variable-frequency carrier oscillator. The phase shift is essential since a two-phase motor is used, and the two currents in the windings must differ by 90° to operate the motor.

If both signals that appear at the balanced modulators are identical in frequency, no mixing process occurs in the balanced-modulator circuitry and, hence, no ac voltage is generated to actuate the motor. When, however, the carrier oscillator drifts slightly, the drift is felt in the final divided signal at the modulator. Now there is a difference in the two signals at the balanced modulator, producing an output ac signal that causes motor armature rotation and a retuning of the variable-frequency oscillator to bring it back to the desired frequency.

While the variable-frequency carrier oscillator is under control of the modulator-motor system, its drift will not be too far from center frequency before correction is made. Thus the difference-frequency signal output from the balanced modulator will always have a low frequency. In many such systems the motor does not respond to signals with frequencies above 1000 Hz, though it rotates for any signal having a frequency below this figure. Thus, a fixed-frequency crystal oscillator is instrumental in maintaining a drift-free carrier frequency in a variable-frequency oscillator.

Another method for minimizing frequency drift of the carrier oscillator is shown in Fig. 7-3. Here, a reactance circuit replaces the motor control. In contrast to the reactance circuit used in the motor-control system, in Fig. 7-3 it serves the dual function of modulation plus drift control.

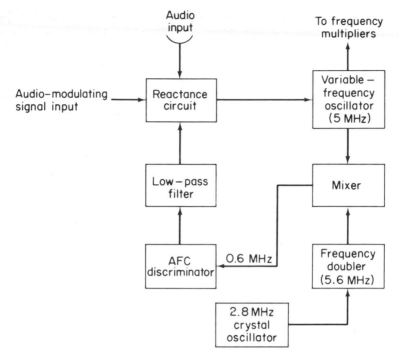

figure 7-3 reactance-circuit direct FM system

Since reactance circuits are not capable of maintaining a linear carrier swing over wide excursions of the resting frequency, it is necessary to keep within the limits of the linear frequency deviation permitted by the particular reactance system utilized. For the system shown in Fig. 7-3, the variable-frequency carrier oscillator is tuned to about $\frac{1}{18}$ of the frequency finally used. Hence, if the final carrier signal frequency is 90 MHz, the carrier oscillator is set at 5 MHz. Maximum deviation is then held at less than 4.2 kHz to maintain a linear frequency shift during the modulation process. If the carrier deviates 4 kHz, the frequency deviation prevailing after the frequency has been raised will be 72 kHz, since the multipliers have increased the deviation by a factor of 18.

Again, a crystal oscillator is used for stability-control purposes. The crystal-oscillator signal frequency is doubled and injected into a mixer, as shown, in conjunction with the 5-MHz variable-frequency oscillator. The mixer circuit functions as a conventional heterodyne mixer circuit used in radio and television receivers. Thus, the signal of the variable-frequency oscillator is heterodyned with that from the crystal oscillator. The mixing process produces a *difference frequency* of 600 kHz (0.6 MHz), and this signal is applied to a automatic-frequency-control (AFC) discriminator circuit, functionally identical to the FM discriminator-type detector described in detail later. Briefly, the discriminator resonant circuits are tuned to the 0.6-MHz frequency signal and no output signal is developed *unless* the signal from the mixer deviates from this resonant frequency.

Thus, if the variable-frequency oscillator drifts from the 5-MHz setting, the difference-frequency signal procured from the mixer will no longer coincide with the resonant frequency of the discriminator. Consequently, an output voltage is obtained from the discriminator that acts as a control signal for correcting the oscillator drift. The control signal is passed through a low-pass filter and applied to the reactance circuit, which in turn corrects the frequency drift of the carrier oscillator.

The low-pass filter passes signals having frequencies only up to 10 Hz and thus keeps out the audio-modulation signals contained in the 0.6-MHz signals originally derived from the modulated carrier oscillator. Tle elimination of the audio signals here prevents their influence on the control system. If the audio components were not filtered out, they would create reactive components opposite to those created by the audio input from the modulating system, thus tending to nullify the frequency deviations that constitute the carrier modulation.

Since the variable-frequency oscillator drift is at a very slow rate, the voltage change at the output of the discriminator changes at a rate very much less than the 10-Hz limit of the low-pass filter circuit. Hence, only the slowly changing correction voltages are applied to the reactance circuit for frequency-control purposes. A variation of the system shown in Fig. 7-3 is given later for the discussion of the control discriminator circuit (Fig. 7-8).

7-3. *indirect frequency modulation*

As described in Chapter 2, the number of *significant* sidebands produced during the frequency-modulation process depends on the maximum carrier swing and the frequency of the audio signal utilized to produce the deviation. As the deviation is reduced, fewer significant sidebands are developed. *At very small carrier deviations, when the shift is less than approximately 30°, only two significant sidebands will be produced* in FM. Thus, at low modulation levels the number of sidebands produced in FM equals those for AM.

The difference, however, is that the two sidebands produced during low levels of frequency modulation are displaced 90° with respect to the carrier, as opposed to AM in which the two sidebands are in phase relationship to the carrier. (See Section 2-4.)

This similarity between AM and FM means that AM can be converted to FM by displacing the two sidebands that are produced by 90° and combining them with the carrier again. By shifting the phase of the sideband signals, the phase modulation that results can be utilized for frequency-modulation transmission. The system is termed *indirect frequency modulation,* since the carrier is not made to shift directly by the audio signals.

The low deviation present when only two sidebands are produced in FM can be increased by increasing the carrier frequency. When frequency-multiplier stages are used, they not only increase the frequency of the carrier signal, but also the extent of deviation. For any instant in deviation, a specific frequency exists, and this, in turn, will be doubled or tripled just as with the carrier signal.

The basic indirect-frequency-modulation system is shown in Fig. 7-4.

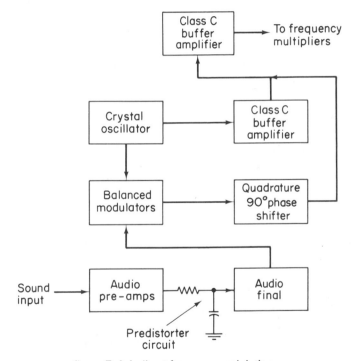

figure 7-4 indirect frequency molulation

Again a crystal oscillator is used for stability, and its signal, after frequency multiplication, becomes the carrier frequency. The amplified audio-signal output from the final audio stage is applied to a balanced modulator in conjunction with the signal from the crystal oscillator. In the balanced modulator the audio signals amplitude-modulate the carrier signal, producing two sideband signals that are applied to a 90° phase-shifting (quadrature) circuit, as shown. The two sidebands, which have been shifted 90°, are then combined with the carrier signal at the output of the buffer amplifier that follows the oscillator. Thus, indirect frequency modulation is obtained. Subsequent stages raise the carrier frequency, as well as the deviation, to that required.

The balanced modulator suppresses the carrier so that the output consists only of the sideband signals. The circuit details are discussed more fully later and illustrated in Fig. 7-7.

In the direct FM discussed earlier, the audio-signal *frequency* determines the *rate* at which the carrier deviates during modulation. Thus, if the audio signal has a frequency of 2000 Hz, the carrier deviates at a 2000-Hz rate on each side of its center (resting) frequency. In direct FM, when the audio-signal *amplitude* is increased, the *extent* of the carrier deviation also increases, though the rate of carrier shift changes only for a change in the frequency of the audio signal. With indirect FM, however, the phase shift of the amplitude-modulated sidebands results in *phase modulation*. (See Chapter 2.) Such phase modulation, while related to frequency modulation, has some characteristics that differ sufficiently to necessitate correction circuitry.

The dissimilarity exists because, during phase modulation, the deviation of the carrier is a function of the frequency of the audio-modulating signal multiplied by the maximum phase shift permitted. Thus, higher-frequency audio signals cause a greater swing of the carrier signal (deviation) than lower-frequency audio signals. This is in contrast to frequency modulation in which only the amplitude of the audio component influences the amount of carrier deviation. (Refer again to Sections 2-3 and 2-4.)

To equalize the deviation so it conforms to that prevailing for FM, a correction circuit is employed, as shown in Fig. 7-4. This resistor–capacitor combination is referred to as a *predistorter*. The series resistor has a high ohmic value compared to the reactance of the shunt capacitor for the full range of audio-signal frequencies used. Thus, the audio signals impressed across the predistorter undergo no appreciable phase change between the signal voltage and signal current.

The audio-signal output from the predistorter is obtained from across the shunt capacitor; hence, the amplitude will vary for signals of different frequencies. Signals of low frequency encounter a high capacitive reactance and little shunting occurs. Hence, virtually full signal amplitude is applied to the stage following the predistorter. As the frequencies of the signals increase, however, the capacitive reactance decreases; hence, the capacitor has a greater shunting effect.

Thus, the predistorter applies signals of decreasing amplitude to the audio stage for increasingly higher audio-frequency signals. Since the phase-modulation process causes a greater carrier frequency deviation for the higher-frequency audio signals, the predistorter, having the opposite effect, nullifies this characteristic of phase modulation and converts the process to the equivalent of frequency modulation. Thus, all audio signals of the same amplitude produce the same amount of carrier deviation, independent of the frequency of such audio signals.

7-4.   *mixing–combining circuits*

Circuits that accept two or more signals and blend these together to form a composite signal are termed mixing circuits, or combining circuits. On occasion the term *adder* will be used to indicate the basic function. Mixers that have nonlinear characteristics produce sum and *difference* signals in addition to the original signals, as was described in the latter sections of Chapter 6. If circuitry has substantially linear characteristics, distortion will be at a minimum and only the combined original signals will appear at the output.

Several mixing circuits are shown in Fig. 7-5. Figure 7-5a shows how two signals may be applied to a single base-input section of a transistor. Resistors

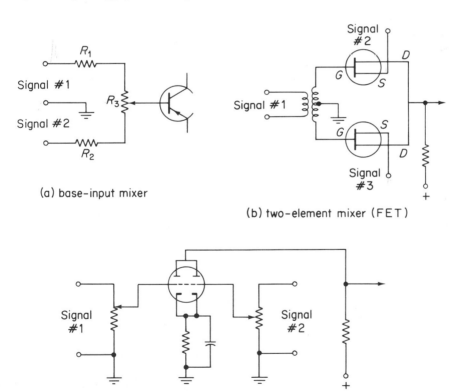

(a) base-input mixer

(b) two-element mixer (FET)

(c) two-element mixer (double triode)

figure 7-5 mixing circuits

R_1 and R_2 are for isolation purposes and minimize interaction and coupling between the two input systems. Resistor R_3 is a balance control for adjusting the relative amplitudes between the two signals.

Less interaction between signal-source circuitry is obtained by applying the input signals to different elements of the transistors or tubes. Figure 7-5b illustrates this method. By using two FET units (or two transistors) with the output elements connected together for a single output, several inputs are provided. For the one shown, however, signal 1 is applied to the input gate elements by a tapped-secondary transformer, providing out-of-phase signals at the gates. The other two signals are applied to the source elements as shown. Such a system is useful for carrier suppression after modulation, as more fully described later. While signal 1 can undergo interaction between the other signals within the FET units, the output for signal 1 is eliminated, since the out-of-phase conditions at the gates cause a current increase in one FET and a corresponding current decrease in the other. Since the total drain current flows through the single output resistor, the opposing effects of the currents are nullified.

Figure 7-5c shows a common form for combining two signals in a dual-triode vacuum tube. The individual signals are applied to separate grid circuits, thus effectively isolating the two and preventing interaction between them. Since each grid is influential in regulating current-flow amplitudes within the tube, their combined results are obtainable from the output by tying the plates together and using a single output resistor, as shown. The cathode resistor sets up a potential difference between cathode and ground and makes the grid more negative than the cathode, thus supplying the necessary grid bias. The shunting capacitor provides a low reactance for signal voltages and thus minimizes their effect across the cathode resistor.

7-5.   varactor and reactance circuitry

Even before the advent of solid-state devices, the vacuum tube was used in circuitry to provide the characteristics of inductance or capacitance without using an actual coil or capacitor. Thus, it provided a convenient method for frequency control and frequency modulation by eliminating mechanical devices. Transistors and FET units can be substituted for tubes to provide for the same type of *reactance* circuitry.

A solid-state device particularly suited to control and frequency-modulation purposes is the *varactor diode*. This unit is a *p–n* junction device that takes advantage of the voltage-variable depletion capacitance of the back-biased junction. Its usefulness extends to tuning, switching, limiting, and pulse shaping or shifting. The nonlinear characteristics of such diodes also aid in harmonic-signal generation. For variable-capacitor use, two diodes are usually connected back-to-back to minimize distortion.

A typical circuit using varactor diodes for frequency control of an oscillator signal is shown in Fig. 7-6a. The two reactive diodes shunt the oscillator inductor and replace the mechanical variable capacitors. As the reverse-bias voltage applied to the junction of the diodes is varied, the capacitance value changes and thus alters the frequency of the signal generated. A filter network is usually employed for correction of any undesirable nonlinearity.

Figure 7-6b shows the basic circuit of a reactance simulator using a vacuum tube. The anode and cathode of the reactance tube are coupled to the resonant tank circuit of the RF oscillator as shown, with C_2 blocking the direct current and preventing a short to ground. Thus, the *signal from the oscillator* appears across the series network composed of C_1 and R_1. The capacitance value of C_1 is selected so that the reactance of C_1 is approximately 10 times higher in value than the ohmic value of R_1. Consequently, current for the C_1–R_1 network *leads* the *signal voltage* obtained from the oscillator. However, since the signal applied to the grid of the tube is obtained from across R_1 only, the voltage *at the grid* will be in phase with the C_1–R_1 network current, since there is no phase shift across a pure resistance. Because the signal current through the tube is in phase with the voltage at the grid, the plate current will lead the oscillator signal voltage by 90°. The phase relationships are shown in the vector drawing of Fig. 7-6c, where E_o is the oscillator signal voltage, I_p the plate current, E_{R_1} the voltage across the resistor, and I_{CR} the network current.

Since a capacitance has a leading current, the reactance-tube circuit simulates a capacitor and, hence, has reactive characteristics. Thus, the coupling of the reactance circuit across the oscillator's resonant circuit introduces a specific amount of shunt capacitance that affects tuning. If an audio signal is applied to the input of the reactance tube, it will cause the frequency of the oscillator to vary at a rate depending on the frequency of the audio voltages. If the audio-signal voltage is increased in amplitude, it will result in a greater oscillator-frequency change. Since reactance $(X_C) = E/I$, a change of plate current alters the amount of capacitive reactance and, hence, the amount of capacitance.

The circuit of Fig. 7-6b can be made inductive by interchanging the placement of R_1 and C_1. When this is done, however, an additional capacitor will have to be included in series with the resistor between the grid and plate to avoid direct coupling of supply voltages. When C_1 and R_1 are interchanged, the ohmic value of R_1 is increased to 10 times the value of the reactance of C_1.

This reactance-circuit principle can also be applied to transistors, including the FET unit, as shown in Fig. 7-6d. The operational theory is

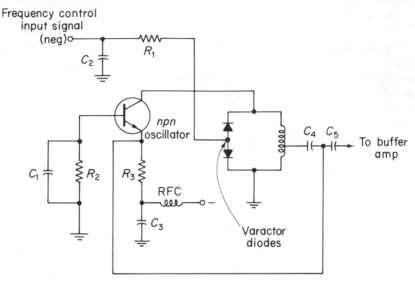

(a) frequency control using varactor diodes

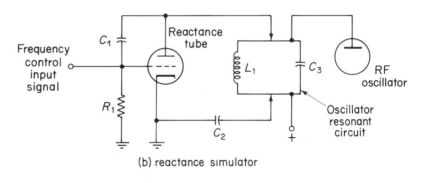

(b) reactance simulator

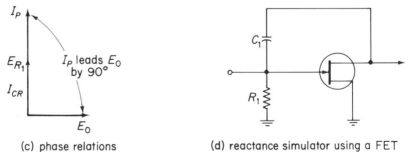

(c) phase relations

(d) reactance simulator using a FET

figure 7-6 varactor and reactance circuitry

identical to that given for the reactance-tube circuit. For all these circuits, a dc voltage can also be used to control frequency drift, as shown in Fig. 7-3.

7-6. balanced modulators

The balanced modulators shown in the block diagram of Fig. 7-4 serve the dual function of modulating the carrier to produce sidebands and then suppressing the carrier. Thus, only the sideband signals are obtained from the output of the balanced modulators.

Two versions of solid-state balanced modulators are shown in Fig. 7-7. The one using *pnp* transistors (Fig. 7-7a) could, of course, also be designed for

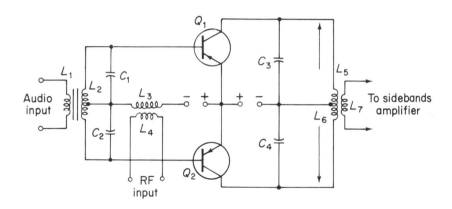

(a) transistor modulator

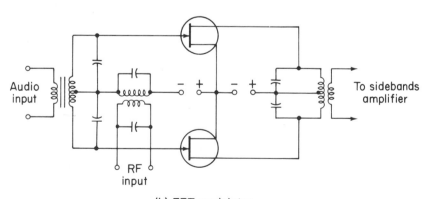

(b) FET modulator

figure 7-7 balanced modulators

npn transistor use by appropriate battery polarity reversals. Note that the RF carrier signal is injected in series with the supply potential and center tap of the input transformer L_2. Consequently, the RF input signal is applied in phase to both base elements of transistors Q_1 and Q_2. Thus, a single RF signal alternation across L_3 causes both base elements to have the same forward-bias change. Hence, if the voltage across L_3 opposes the negative forward bias, the reduction in bias causes a decrease in the current in both collector-emitter circuits. Since the collectors of the balanced modulator are connected in push–pull, electron flow through each transistor is in the direction shown by the arrows. Thus, current changes in L_5 and L_6 are equal and opposite to each other; consequently, cancellation occurs for such a current change representative of the RF signal. (It is assumed that the circuit is closely balanced, with matched characteristics for Q_1 and Q_2.)

The audio signals appearing across the secondary of the input transformer (L_2) are applied to the base inputs of the two transistors in typical 180° phase difference characteristics found in push–pull circuitry. Thus, at the input the signals have voltage relationships as follows for the carrier (E_c) and the modulating voltage (E_m):

$$E_{b_1} = E_c \cos \omega t + E_m \cos \omega t$$
$$E_{b_2} = E_c \cos \omega t - E_m \cos \omega t \tag{7-1}$$

Since the audio signals cause changes in collector current flow, the carrier-frequency currents within each transistor undergo modulation. Carrier modulation produces sidebands, and these find resonant circuits in the output consisting of C_3 and L_5 for Q_1, and C_4 and L_6 for Q_2. These resonant circuits have a low impedance for the audio signals and minimize their appearance at the output. Since the carrier has been suppressed, only sideband signal energy is obtained from the balanced-modulation system. Capacitors C_1 and C_2 at the input have a low reactance for the RF signal energy and, hence, provide coupling to the base inputs. For the audio appearing across L_2, however, these capacitors have a very high reactance and, hence, offer no shunting effect.

For the balanced modulator shown in Fig. 7-7b, the theory of operation is the same, the only difference being the use of FET units. Potentials and polarities for the drain and source elements must, of course, be applied as indicated. Also, as shown, capacitors may shunt input inductors to form resonance for the RF signals.

7-7. discriminator control system

As shown in Fig. 7-3, an automatic-frequency-control (AFC) discriminator is used to keep the frequency-modulated oscillator stable around the assigned carrier frequency, while at the same time permitting the oscillator frequency to be shifted in accordance with the modulating signals. Details of the control discriminator are shown in Fig. 7-8, plus illustration of another version of frequency multiplication to achieve desired results.

Inductor L_2 has a reference voltage drop across it obtained from the resonant circuit of the mixer via the coupling capacitor C_6. Inductor L_1 is coupled to L_3 and L_4 by conventional transformer arrangement. Thus, the output of the mixer is developed across the secondary, L_3 and L_4, while a reference voltage is established across L_2. This sets up a voltage distribution

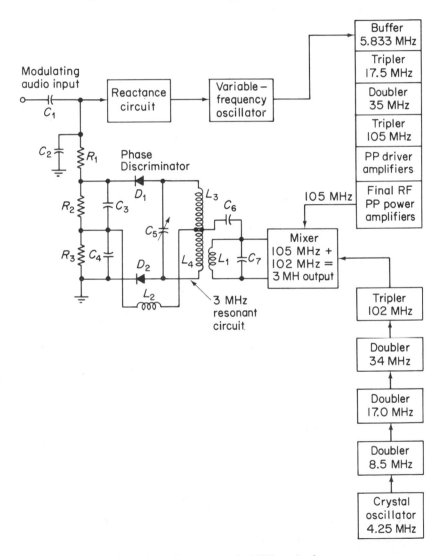

figure 7-8 discriminator control of FM carrier frequency

having vector relationships similar to those described for the FM discriminator detector described later in this chapter and to which reference should be made for a thorough analysis of signals and phases.

The phase discriminator develops a correction output voltage that is applied to a reactance circuit, which, in turn, controls the frequency stability of the variable-frequency oscillator. As shown, a crystal oscillator is used as the frequency-reference circuit, and this oscillator may be operated at a much lower frequency than the final FM carrier to permit use of a larger crystal for convenience in temperature control and efficiency in signal output.

Thus, if the crystal-oscillator signal frequency is selected as 4.25 MHz, a series of doublers and triplers raises the frequency until it is near that for the signal from the final RF amplifiers, as shown in Fig. 7-8. For this particular transmitter, the carrier frequency is 105 MHz and, hence, it is within the standard-broadcast FM band of 88 to 108 MHz. This 105-MHz signal represents the multiplied frequency from the frequency-modulated oscillator, which has been set at a low value so that the small deviation established by this modulation system may be increased to that required during broadcast. This is advisable because of the limited capacitance variation of reactance circuits for a linear deviation. Thus, frequency deviations from a reactance circuit may be only a few kilohertz and must be increased for full deviation by frequency multiplication of the modulated carrier.

Thus, if the variable-frequency oscillator is set for 5.833 MHz, a series of doublers and triplers finally raises the carrier frequency to 105 MHz for a particular station. If the deviation is 4 kHz, this would be tripled, doubled, and again tripled to reach a value of approximately 72 to 75 kHz for maximum permissible deviation for this FM band.

The signal from the RF final amplifier is combined in the mixer circuit with the output signal of the final multiplier from the crystal oscillator. For Fig. 7-8, this consists of 105 plus 102 MHz, which heterodyne together to produce a difference frequency of 3 MHz. This 3-MHz signal is impressed on the phase discriminator, which is tuned to this resonant frequency.

Thus, if the signals from the RF final amplifier and the final tripler from the crystal oscillator remain the same, the output frequency from the mixer holds at 3 MHz, and the voltage at the output of the phase discriminator (across R_2 and R_3) is zero. Thus, no correction bias is sent to the reactance circuit. The phase discriminator has zero output because the individual voltage drops across R_2 and R_3 will be equal but opposite in polarity, thus canceling out, as discussed more fully later for the FM discriminator detector.

If the variable-frequency oscillator drifts from its assigned frequency, the mixing process would no longer produce 3 MHz, and one diode of the discriminator would conduct more than the other, producing unequal voltage drops across the output resistors, thus developing a correction voltage. Whether the signal frequency from the mixer rises above 3 MHz or falls below it determines whether a positive or negative potential is developed at the discriminator output. The output-signal polarity, in turn, determines whether the reactance circuit increases the frequency of the variable oscillator or decreases the frequency of the output signal.

Capacitor C_2 has a low reactance for RF signals and, hence, filters them from the output of the discriminator. This prevents their entry into the reactance tube, since the frequency-modulation components present in the phase discriminator would affect the performance of the reactance circuit.

The reactance of C_2 is sufficiently low for RF filtering purposes, yet does not shunt the dc voltage developed by the discriminator, and they appear at the reactance circuit. The variable oscillator will not drift rapidly unless it has developed some defect in a circuit component. Thus, a slow-rate drift of the variable oscillator develops a correction voltage, while rapid deviations caused by the modulation process are effectively bypassed by the filter network composed of R_1 and C_2.

7-8. stereo FM multiplex system

As mentioned in Section 7-1, in addition to mono broadcasting, some FM stations offer other types of transmissions, such as the multiplex stereo or SCA system. The basic stereophonic FM multiplex system is shown in block-diagram form in Fig. 7-9. The audio signals picked up by microphones, or from records or tape, etc., have a frequency response *of at least* 50 Hz to 15 kHz, with a 75-μs pre-emphasis for each channel, as described more fully later.

Stereo broadcasting is not simply a matter of modulating two separate carriers, one for the left sound and the other for the right, since spectrum space must be conserved. Also, the stereo broadcasts must be compatible; that is, a mono receiver must be able to pick them up (as, also, the stereo receiver must be capable of mono reception). Hence, special mixing of left- and right-channel sounds must precede usage of balanced modulators for stereo carrier suppression, as shown.

As discussed in Chapter 2, an AM signal can be overmodulated and cause distortion, while for FM there are no limits to the *extent* or *rate* of carrier deviation. For standard FM broadcasting, however, 100 per cent modulation is defined as a carrier deviation of 75 kHz each side of its resting frequency (see Fig. 2-10). In stereo or other multiplex systems we must still stay within the assigned limits defined as 100 per cent modulation; hence, the various modulating signals must share the modulation process without exceeding 100 per cent.

Modulating-signal frequencies can extend above the specified 15 kHz required for good fidelity. For such higher modulating-signal frequencies,

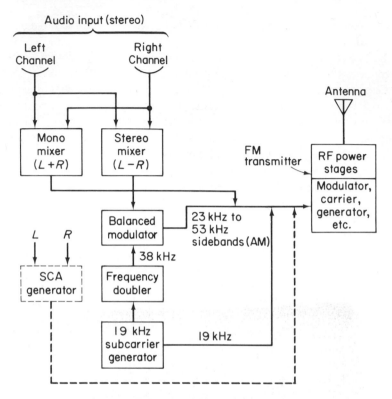

figure 7-9 stereo FM multiplex system

the *rate* of carrier deviation increases, though the *extent* of carrier excursions each side of center frequency still depends on modulating-signal amplitudes. Thus, in FM multiplex systems, signals having frequencies above 15 kHz are utilized for modulation purposes.

For compatibility the stereo station transmits a mono signal by adding the two signals received from separate sources. This addition $(L + R)$ is fed to the modulating section of the main FM transmitter and forms what is commonly referred to as the *main channel*. As shown in Fig. 7-10, the mono modulating signal, with a minimum frequency range of 50 Hz to 15 kHz, occupies the initial portion of the complete modulating-signal spectrum required for multiplexing.

For stereo an additional signal is required for obtaining the distinct left and right channels during reception. To multiplex this signal, a *difference signal* is formed by subtracting the R signal from the L (feeding both L and R to a mixer circuit, with R 180° out of phase). This difference signal modulates an additional carrier (called a *subcarrier*) and the AM process forms sidebands. These sidebands, in turn, share in modulating the transmitted FM carrier, as shown in Fig. 7-9. The subcarrier is suppressed with circuitry

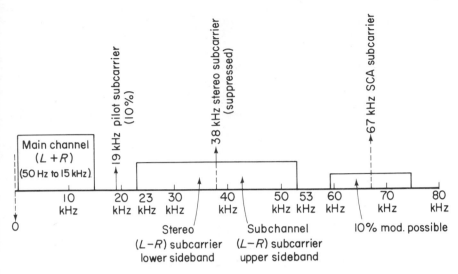

figure 7-10 spectrum of modulation signals for FM station with stereo
and SCA

described in Section 7-6. This missing subcarrier must be reinstated in the
receiver for reception of the stereo signal.

The subcarrier's frequency is 38 kHz and is obtained by doubling a 19-
kHz signal obtained from the generator, as shown in Fig. 7-9. This 19-kHz
signal is transmitted (by modulating the carrier) for synchronization of the
stereo detection in the receiver. This 19-kHz signal is known as the *pilot*
subcarrier and is held to ± 2 Hz of assigned frequency. It modulates the transmitted FM carrier by only 10 per cent, but this is sufficient to enable the
receiver to double the frequency of the pilot signal to obtain the 38-kHz
subcarrier signal for reinsertion with the stereo sidebands, as more fully
detailed in Section 7-14.

The sidebands obtained by modulating the 38-kHz subcarrier with the
$L - R$ signals are situated above the mono modulating signals, as shown in
Fig. 7-10. These sidebands thus occupy a region between 23 and 53 kHz.
Note that the lower $L - R$ sidebands (as well as the upper stereo $L - R$
sidebands) occupy the same bandwidth of 15 kHz as the $L + R$ mono signals.
As with the mono signals, the audio-signal frequency range is at least 50 Hz
to 15 kHz. Thus, the entire *multiplex modulating signal* for stereo FM transmission is composed of the mono $L + R$ signal in the audio range (50 Hz
to 15 kHz), a supersonic 19-kHz pilot subcarrier signal, plus the $L - R$
supersonic signal (23 to 53 kHz) with suppressed 38-kHz carrier. With SCA
(discussed later) the complete modulating-signal spectrum is as shown in
Fig. 7-10.

In stereo broadcasting the subcarrier (19-kHz pilot signal) is permitted a
10 per cent injection, as mentioned earlier, thus providing a carrier-frequency

deviation of 7.5 kHz each side of resting frequency. If no SCA transmission is included, there remains then 90 per cent possible modulation for the other channel signals during multiplex broadcasting. Hence, the remaining 90 per cent modulation capability must be divided between the mono and the stereo channels. If the mono signal modulates the main channel 90 per cent, the stereo sidebands would not modulate the FM carrier at all. A seesaw effect is usually present for L- and R-channel amplitudes, since, during stereo broadcasting, amplitude differences usually exist between left and right program material.

Thus, the 90 per cent remaining modulation possibility is shared by the $L + R$ and the $L - R$ signals, and it is not possible for both to modulate the transmitter by 90 per cent under normal and proper operating conditions. The $L + R$ can, however, modulate the main channel by 90 per cent (during mono broadcast only), while the $L - R$ sidebands are capable of only up to 45 per cent modulation (and this maximum value only holds for a single left signal or a single right signal). If the stereo sidebands are modulating at the 45 per cent level, the $L + R$ modulating maximum does not exceed this 45 per cent either.

7-9. the SCA multiplex system

The SCA (Subsidiary Communications Authorization) system permits an FM station to carry another broadcasting channel in addition to the standard FM transmission. When the latter is for general-public use, the SCA is for private subscribers who pay a fee for background music in stores, doctors' offices, etc.

To multiplex the SCA transmissions, an SCA generator is used, as shown in the broken lines in Fig. 7-9. This unit is essentially a complete miniature FM transmitter (compared to the main transmitter), with a center subcarrier frequency of 67 kHz. (The FCC does not require the specific 67-kHz subcarrier frequency, though this value has been virtually standard for stereo–SCA combination stations.)

If a station's main channel is mono and the SCA transmission is also used, the 67-kHz signal would be applied to the transmitting section along with the conventional modulating signals. Since a 75-kHz deviation each side of the carrier is considered 100 per cent modulation, the limits thus imposed must not be exceeded when the SCA signal is injected. Hence, the modulation level for standard FM is reduced to accommodate the SCA signals. The FCC limits the SCA injection to 30 per cent of total modulation. Thus, for a mono station also using SCA, the main channel modulation is set at 70 per cent (52.5 kHz each side of resting frequency).

Within the SCA generator the audio *frequency-modulates* the 67-kHz subcarrier for a maximum deviation usually held to 7.5 kHz each side of center, with a frequency response range of 30 to 7500 Hz and a deviation ratio of 1. Thus, when the SCA system is multiplexed with standard frequency modulation, the system is sometimes referred to as *FM-on-FM*. Obviously the system suffers somewhat in terms of dynamic audio range and frequency response because of the channel sharing. Also, because of the low deviation ratio, the signal-to-noise ratio suffers, and some station muting is undertaken on occasion to silence noises between audio modulations.

As shown earlier in Figs. 7-9 and 7-10, the SCA system is feasible also for multiplexing with stereo. With stereo, however, the SCA injection is held to 10 per cent to prevent exceeding frequency standards. Thus, the SCA signal causes a carrier deviation of only 7.5 kHz each side of center frequency. Hence, the 10 per cent injection for SCA plus the 10 per cent for the pilot subcarrier reduces the remaining modulation capability to 80 per cent only.

This 80 per cent modulation capability is the maximum for the combined main-channel modulation *and* the stereo sidebands. Thus, without the stereo-signal modulation, the main-channel modulation is 80 per cent and virtually monophonic. With L at maximum and R at zero, the main channel modulation would be 40 per cent, the same as for the sidebands. (The stereo sidebands, during proper operation, would never modulate more than 40 per cent.) With reduced modulation, the deviation ratios for the various main channel and subcarrier FM also change, with less sideband power developed during transmission. [See Eqs. (2-6) and (2-7).]

7-10. pre-emphasis and de-emphasis

In the mono, stereo, or SCA frequency-modulation systems a noise-reduction process is used involving a system of *pre-emphasis* at the transmitter and *de-emphasis* at the receiver. Interelement transistor noises, tube noises, and circuit noises are generated at a fixed-amplitude level in a given system. Hence, the signal-to-noise ratio can be increased by raising the level of the signal over the constant-level noise signal. Since the noise generated increases for higher-frequency audio signals, it is necessary to raise the level of the audio-frequency signals at an increasing rate for higher-frequency signals, thus employing a system of pre-emphasis.

The FCC has set the rate of incline for the pre-emphasis process, and the rise in the amplitude of the audio frequency signals starts at about 400 Hz and rises gradually; at 1000 Hz the increase is 1 decibel (dB); at 1500 Hz the increase is almost 2 dB. At 2000 Hz, there is an amplitude rise of approximately 3 dB; at 2500 Hz there is almost a 4-dB increase. From this point on, the increase in pre-emphasis is virtually linear, reaching 8 dB at 5 kHz and 17 dB at 15 kHz.

The basis circuit for pre-emphasis is shown in Fig. 7-11a. This resembles the standard input circuit of an amplifier stage, with C_1 as the coupling capacitor and R_1 the base resistor across which the signal appears. Instead of having a large-capacitance value for C_1, however, a reduced capacitance is used to provide the necessary rise in amplification for the higher-frequency audio signals. The time constant (RC) for the pre-emphasis network is 75 μs, as established by the FCC. That is, $RC = 75 \times 10^{-6}$. This gives optimum results without excessively increasing frequency deviation by virtue of the increase of signal amplitude.

At the receiver it is necessary to de-emphasize the gradual rise in signal frequencies to prevent shrill and harsh high-frequency audio response. A typical circuit is shown in Fig. 7-11b and consists of the series resistor R_1 plus the shunting capacitor C_1. Such a de-emphasis network usually follows the FM detector and often precedes the volume control, as shown. The time constant is again 75 μs, and higher-frequency signals find an increasingly lower reactance for C_1, with consequent decrease in amplitude because of the shunting effect.

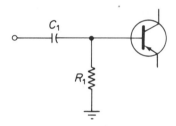

(a) basic circuit for pre-emphasis

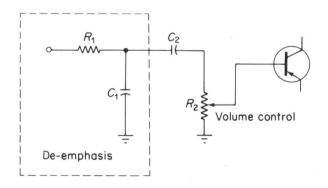

(b) de-emphasis circuitry

figure 7-11 pre-emphasis and de-emphasis

7-11. FM discriminator detector

As described in Section 6-4, the diode is ideally suited for demodulation purposes since it has rectifying characteristics. Thus, the same unit is used in FM detection, except the circuit must be so designed that the audio components obtained are the result of demodulating carrier-frequency deviations rather than demodulating amplitude changes.

The discriminator FM detector shown in Fig. 7-12 uses two diodes in a

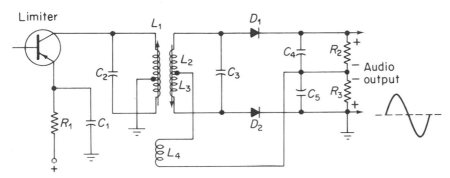

figure 7-12 FM discriminator detector

symmetrical demodulation system, with the audio components developed across the two output resistors R_2 and R_3. Since, however, the discriminator detector not only demodulates FM, but also AM, it is necessary to eliminate any amplitude changes (such as static or other types of amplitude distortion) that may be present on the frequency-modulated carrier arriving at the tuner of the receiver. Hence, the discriminator is preceded by a *limiter* stage for removal of unwanted AM.

As with the AM superheterodyne principle described in Section 6-7, the incoming FM carrier is heterodyned with the signal generated by the local oscillator of the tuner to produce an IF signal. As with AM, one particular intermediate frequency seems to be better suited than others in terms of elimination of interference, and with general-public FM (88 to 108 MHz), the IF signal that has become standard has a frequency of 10.7 MHz. Other FM services (such as the sound accompanying the video signal in television broadcasting) have intermediate frequencies to suit the particular carrier frequencies used.

As explained in Chapter 3, clippers of various types can be used to remove undesired portions of waveforms from an existing signal. By overdriving an amplifier (Fig. 3-22b) waveform peaks can be eliminated for both positive and negative alternations, thus obtaining a constant-amplitude output. When a resonant circuit is present, as in Fig. 7-12, the transistor can be biased at or slightly beyond the cutoff point and sufficient signal drive applied to the base input circuit to cause saturation for every alternation that permits transistor collector current to flow. Thus, collector current flows from zero to saturation and, hence, is essentially pulsating direct current. However, the flywheel effect of the resonant circuit composed of C_2 and L_1 reproduces the sinewave characteristics of the carrier with a constant amplitude, thus eliminating amplitude-modulation components.

The signal from the limiter is transferred to the discriminator by the transformer action between L_1 and the secondary L_2 and L_3. Note, however, that L_4 is coupled to L_1 and, hence, picks up a reference signal for purposes, described later in this section. Such a reference voltage may also be obtained by capacitive coupling between the limiter output and the discriminator.

Diodes D_1 and D_2 rectify the signal energy obtained from the secondary and cause current flow through resistors R_2 and R_3. The resultant voltage drops have polarities as shown in Fig. 7-12. Note that opposing polarities exist between R_2 and R_3, providing zero output if the voltage amplitude across R_2 equals that across R_3.

The function of the discriminator depends on the voltage relationships of the signals developed across the four transformer sections between the two stages. As shown in Fig. 7-13a, phase differences exist between the primary signal voltage E_p, the induced voltage across the secondary E_{ind}, and the primary current I_p. Since L_1 has inductive characteristics, primary current I_p lags the primary voltage E_p. When, however, the primary current undergoes its greatest *change* (positive peak to negative peak), the resultant lines of force induce a voltage across the secondary E_{ind}.

As shown in Fig. 7-13a, the induced voltage lags primary current by 90°. Thus, there is a phase difference of 180° between the *primary voltage* and the *induced voltage*. Inductor L_4, which also picks up an induced voltage from L_1, has transposed leads so that the signal voltage across L_4 is in phase with that across L_1, thus also causing the signal voltage across L_4 to be 180° out of phase with E_{ind}.

Voltage relationships in phasor form are shown in Fig. 7-13b. Note that the signal voltage across L_4 (E_{L_4}) is represented by a vertical arrow pointing upward to indicate an out-of-phase relationship with the induced signal voltage E_{ind} (shown with a downward vertical arrow).

As shown in Fig. 7-12, the signal voltage across the secondary also finds a shunt capacitor C_3, which forms a resonant circuit with L_2 and L_3. At resonance the respective reactances of the inductors and capacitor cancel, leaving only pure resistance. Thus, secondary current I_s is in phase with the induced voltage, and such I_s is also represented by an arrow in the same direction as E_{ind}, except it is drawn with a heavier line to emphasize its difference.

Because of the center tap of the secondary, it is divided into two in-

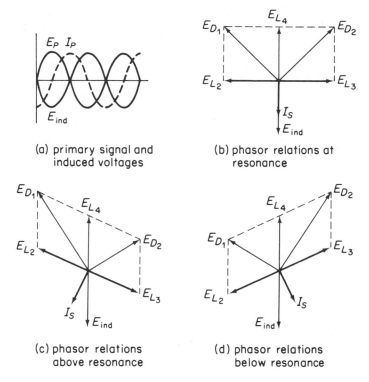

(a) primary signal and
induced voltages

(b) phasor relations at
resonance

(c) phasor relations
above resonance

(d) phasor relations
below resonance

figure 7-13 phasors for FM detection

ductors with individual signal voltages. Thus, voltage across L_2 is applied
to the diode D_1 from the top of L_2 as well as the center tap coupled to the
diode via R_2. Diode D_2 is similarly coupled to L_3 via the center tap and
resistor R_3.

Even though a resonant circuit is involved, the signal voltage across L_2
is out of phase with the secondary current by 90°, since this relationship
prevails for an individual inductance. Also, the signal voltage across L_3
is out of phase with I_s by 90°, as shown with heavy lines in Fig. 7-13b. Signal
voltages across the transformer secondary winding cause D_1 to conduct at
one instant and D_2 at another. Even though the diodes thus conduct alternately, both voltages are, of necessity, represented simultaneously in the
vector diagram.

Because there is a 90° phase difference between the signal voltage of L_4
and the voltage of L_2, a vector representation of the voltage for D_1 must be
drawn with the diagonal arrow shown in Fig. 7-13b. Similarly, a slanting
arrow is shown at the right of the same drawing for the voltage across D_2.
These values complete the vector representation and indicate the *phases*
of the *voltages* when the carrier is at its *center* frequency. The voltage ampli-

tude for each diode is identical; hence, the individual voltage drops across the output resistors R_2 and R_3 will also be the same. Since these are out of phase, no output signal is obtained.

During carrier deviation of the FM process, the shift away from center frequency also means a shift away from the resonant condition prevailing at the resting frequency. Thus, when the carrier shifts to a higher frequency, the resonant circuit formed by the transformer secondary and C_3 of Fig. 7-12 becomes primarily inductive.

The inductive characteristic occurs because signal energy is induced from L_1 into the secondary winding by the magnetic lines of force cutting across the secondary. This induction does not involve the shunt capacitor C_3, and the latter must be charged with energy by virtue of its coupling to the secondary. Hence, the voltage is induced as though into a series-resonant circuit. As resonance, the series circuit has a low impedance, with inductive and capacitive reactances equal though opposite in phase.

When frequency rises above resonance, the inductive reactance rises, but capacitive reactance decreases. Thus, the rise in inductive reactance offers the greatest opposition to signal current and causes the circuit to become primarily inductive. The result is a lag of secondary current I_s, as shown in Fig. 7-13c. Individual voltages across the secondary inductances L_2 and L_3 are still 90° out of phase with the signal current in the secondary because of the inductive characteristics of L_2 and L_3. Thus, the vector arrows for E_{L_2} and E_{L_3} must also be shown tilting at an angle to maintain their right-angle relationship with I_s.

As shown, the parallelogram for the voltage of D_1 shows a rise, while the voltage for D_2 declines. The result is a rise of voltage across R_2 but a drop across R_3, causing the production of a positive signal alternation at the output of the discriminator, as shown. When the carrier shifts back to its center again, the secondary current I_s becomes in phase again with the induced voltage E_{ind}, as it was in Fig. 7-13b, and in consequence the output signal drops back to zero.

When the carrier shifts to a frequency below its center frequency, secondary current leads, because the circuit becomes primarily capacitive. Now voltage and phase relationships are as shown in Fig. 7-13d. For D_2 there is a voltage increase, and a decrease for D_1. Now the voltage drop across R_3 rises, while that across R_2 declines, producing a negative-alternation output. Again, when the carrier returns to its resting frequency the voltage and phase conditions of Fig. 7-13b prevail.

Since the rectification processes of D_1 and D_2 cause current flow in unidirectional alternations of the input RF signal, capacitors C_4 and C_5 shunt the output resistors and filter out the RF ripple component. These capacitors have a low reactance for the RF signal being demodulated, but a high reactance for the audio-signal voltages. The audio output signals are applied to a de-emphasis network, coupling capacitor, and volume control, as shown in Fig. 7-11b.

If the frequency of the carrier deviates more rapidly, the audio cycles produced at the output appear in more rapid sequence, resulting in a *higher frequency* signal output of audio. If the carrier makes greater excursions beyond its midfrequency, it causes greater phase shifting in the discriminator vector representations, producing greater potential differences between the output resistors and a greater amplitude of audio-signal voltages at the output.

7-12. FM ratio detector

The FM ratio detector has an advantage over the discriminator-type demodulator in that no limiter need precede it, because it is insensitive to amplitude modulation. As shown in Fig. 7-14, the ratio detector resembles the discriminator in some aspects, though the diodes conduct simultaneously. Since they are in series with the transformer secondary and the output resistors, voltage drops across R_2 and R_3 have an aiding polarity as shown, thus providing a voltage even in the absence of frequency modulation of the incoming carrier. This direct current can, however, be blocked by coupling capacitor C_7.

The factors that cause one diode to conduct more than the other to produce an audio output are similar to those discussed for the discriminator, and the vector diagrams of Fig. 7-13 apply to the ratio detector also. When the incoming carrier signal is modulated, the frequency deviations cause one diode to conduct more than the other because of the same phase changes and

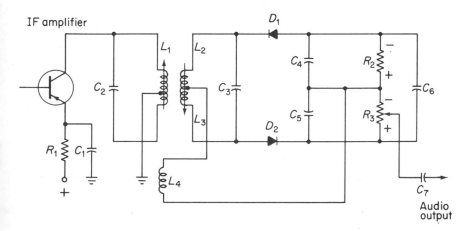

figure 7-14 ratio detector for FM

relationships described for the discriminator. Because of the series arrangement of the diodes, however, differences exist at the output.

For an understanding of output-voltage relationships across R_2 and R_3, assume that current conduction through the diodes and resistors causes a drop of 1 V across each output resistor for a total amplitude of 2 V. Now, if the carrier deviates and D_1 conducts more than D_2, the voltage drop across R_2 may rise to 1.5 V and that across R_3 drop to 0.5 V. Now, the total amplitude across the two resistors is still 2 V.

If the carrier swings in the other direction, D_1 conducts less than D_2, causing the voltage across R_2 to drop to 0.5 V and that across R_3 to rise to 1.5 V. Again, no voltage *change* occurred across the two output resistors. Note, however, that while the *ratio* of voltages across the resistors may change, the total voltage remains the same. Hence, the audio output must be obtained from one of the two output resistors. Since R_3 has one side at ground potential, this resistor is more convenient to obtain the output. By using a potentiometer, the amount of audio obtained can be regulated, thus forming a volume control of one of the ratio resistors.

Capacitor C_6, which shunts the two resistors, has a high capacitance value (usually several microfarads). This capacitor charges to the dc voltage value appearing across the output resistors. Since a capacitance opposes a voltage change, C_6 maintains the voltage across the combination R_2–R_3 resistors at a constant level. Hence, any sudden changes in the total voltage, such as might occur because of sharp static bursts or noise pulses, are minimized.

As with the discriminator, a well-balanced circuit is important for optimum performance. The two diodes should be well matched and the transformer secondary tapped as near the electrical center as possible. Matched resistors and capacitors also help obtain a good balance.

7-13.   gated-beam FM detection

Another FM detector system, the gated-beam demodulator, has much higher sensitivity and output than the discriminator- and ratio-detector processes. Hence, the output is sufficient to drive the audio power output stage of television receivers directly, without having to use one or two stages of additional voltage amplification. Because of its greater sensitivity, a single amplifier stage is usually sufficient to drive the gated-beam detector. Though linearity and fidelity are not equal to the diode-type detectors, the gated beam has been widely used in television receivers in which a common tube often serves as both the sound detector and the sound output, as shown in Fig. 7-15.

The first grid (nearest the cathode) is termed the *limiter* grid, the next grid is the *accelerator*, and the third (nearest the anode) is the *quadrature* grid.

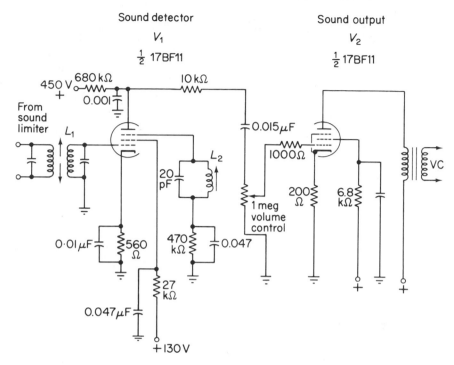

figure 7-15 gated-beam FM demodulator

The accelerator grid structures are, in reality, in the form of plates that help shape the electrons into a narrow beam. As shown, a positive voltage is applied to the accelerator grid, which increases the velocity of the electron beam and forces it through a narrow slot in the accelerator electrode. Since electron flow between the cathode and plate must also pass the limiter and quadrature grids, these two grids also influence current flow.

The limiter grid is sufficiently influential over current flow to cut it off entirely for any negative potential. With zero or positive voltages on the limiter grid, however, current will flow within the tube. If signals in the quadrature circuit (composed of the L_2 inductance plus its shunting capacitor) swing in a negative direction, current within the tube can also be cut off. Hence, both the limiter and quadrature grid can prevent or permit current flow within the gated-beam detector tube.

If the tube bias developed across the cathode resistor network is of low value (around 1 V), an incoming signal has sufficient amplitude to cause the tube to conduct at saturation for positive peaks of the grid signal and at cutoff for the negative peaks of the incoming signal. Because of the design features of the tube, the grid structure releases current flow suddenly and also stops flow very quickly. Hence, current pulses assume square-wave shapes within the tube in the region beyond the input grid. Thus, the tube acts as a self-limiting device and removes amplitude variations in the incoming signal.

The quadrature inductor L_2 and its shunting capacitor form a parallel-resonant circuit tuned to the center carrier frequency of the incoming FM signal. Since this is an IF-type signal, there is no retuning for different stations. For signal input, the space charge (electron cloud) around the cathode structure varies, and the quadrature grid is also influenced by the electron beam because of space-charge coupling. Hence, the square-wave type of signal generated within the tube also appears at the quadrature grid and pulses the quadrature circuit into a resonant flywheel condition.

There is a lag of approximately 90°, however, between the signal voltage across the quadrature circuit and that at the input, because of the nature of the space-charge coupling. Because of this 90° lag, the anode current of the tube is cut off for a longer period of time than would otherwise be the case. The relationship between the limiter grid and quadrature grid influencing the width of plate-current pulses is shown in Fig. 7-16. The plate current can only flow when neither the limiter grid nor the quadrature grid is negative. Thus, only about half of each square-wave alternation reaches the anode during the time the carrier is at its center frequency.

If modulation shifts the incoming carrier higher in frequency, the quadrature circuit shifts to a greater off-resonance condition with respect to the shifted carrier frequency. This causes the quadrature circuit to become predominantly capacitive in its characteristic, since the higher-frequency signal impressed on it will raise inductive reactance and lower capacitive reactance. Since the capacitive reactance is low, it permits greater current flow rather than the higher inductive reactance. The resultant reduction of the parallel resonant circuit's high impedance by the low-value capacitive reactance causes the circuit to be capacitive.

This capacitive characteristic of the quadrature circuit causes signal voltage at the quadrature grid to lag the signal at the limit grid by more than 90° (the prevailing lag when the carrier is at center frequency). Because of the increased difference in phase between the two current-controlling elements, *less* than half of each square-wave alternation arrives at the anode of the tube. Thus, the *average value* of the plate current decreases. When the modulated carrier at the limiter grid shifts lower in frequency, the quadrature circuit again changes, becoming predominantly inductive in characteristic and causing a leading voltage. As shown in Fig. 7-16, more than half of each square-wave alternation now reaches the anode, and this causes an increase in the average value of the plate current.

The 680,000-Ω series resistor and shunt capacitor (0.001) in the anode circuit of Fig. 7-15 form an integration circuit of the type discussed earlier (see Fig. 3-17). Integration produces the average value from the pulse train, and this forms the positive and negative alternations of the audio signals obtained from the demodulation process.

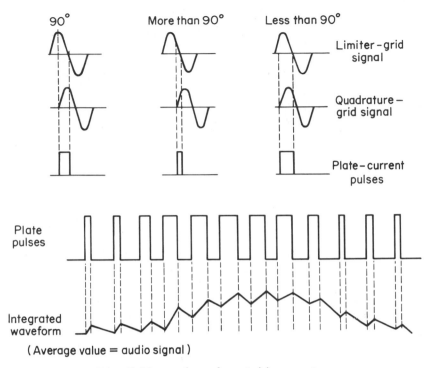

figure 7-16 waveforms for gated-beam system

As shown in Fig. 7-15, conventional circuitry couples the output of the gated-beam detector to the input of the audio power-output stage. A 1-MΩ pot is used for the volume control, and a 1000-Ω series resistor in the grid circuit prevents the latter from being grounded completely when the variable arm of the control is turned to the lowest level. Though both tubes appear as pentodes, the gated-beam detector differs considerably from the pentode design for forming the peculiar characteristics inherent to this demodulation system.

7-14. stereo FM detection

When *stereo* FM transmission was first used commercially shortly after its approval by the FCC in 1961, early receivers and multiplex adapters utilized the matrix-type multiplex decoder system shown in Fig. 7-17. While this particular system has been superseded in modern receivers by a more sophisticated process, it serves as a useful introduction to stereo demodulation principles.

As shown, the initial circuitry up to and including the discriminator is in conventional FM receiver form, though the de-emphasis circuitry is

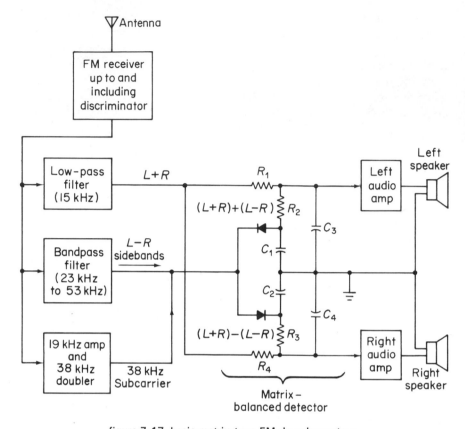

figure 7-17 basic matrix-type FM decoder system

used later in the receiver. Since all the signals making up the stereo broadcast are used to frequency-modulate the carrier (see Fig. 7-9), the FM detector will have an output consisting of the original modulating signals: the $L + R$ signal in the audio range of 50 Hz to 15 kHz, a supersonic 19-kHz pilot subcarrier signal, plus the $L - R$ supersonic signal (23 to 53 kHz with suppressed 38-kHz carrier). If SCA is also used, a 67-kHz carrier is also involved.

All these signals coming from the FM detector comprise what is termed the *composite* FM signal. It must be remembered that the $L - R$ signals are in the form of sidebands, which were derived by amplitude-modulation of the 38-kHz subcarrier (which was eliminated at the transmitter). Even though these sidebands were obtained by the AM process, they were used to *frequency-modulate* the main carrier. Thus, in the receiver these sidebands are obtained by detection in the discriminator (or ratio detector), but they are of a supersonic frequency and inaudible. Initially, they must be recombined with the missing subcarrier of 38 kHz; then the resultant amplitude-modulated signal must be detected in the conventional AM demodulation process.

The composite signal obtained from the discriminator detector feeds three circuits, as shown. The low-pass filter accepts the composite multiplex signal, but passes signals of frequencies only to 15 kHz, comprising the $L + R$ components. Thus, the audio information of the $L + R$ obtained by conventional FM detection is applied to a matrix (adder) circuitry that also has provisions for detection of the $L - R$ components, and includes de-emphasis circuitry for the respective left- and right-channel output signals.

The $L - R$ signals, as discussed in Section 7-8, carry the stereo information. To salvage these signals from the composite signal, a bandpass filter is used with a design that permits passing only of the 23- to 53-kHz $L - R$ sideband signals. (High-pass, bandpass, and other filters are covered in Chapter 10.)

As shown in Fig. 7-17, the composite signal is also applied to circuitry that accepts and amplifies the pilot 19-kHz signal and doubles it to 38 kHz to obtain the carrier that was suppressed at the station. Older receivers used a separate oscillator synchronized by the 19-kHz signal, but recent receivers utilize the doubling method discussed later, which dispenses with the oscillator circuit and the stability-synchronizing problems associated with it. The 38-kHz output signal is added to the $L - R$ signal in the matrix circuit, and the resultant amplitude-modulated carrier is detected by the two diodes shown.

Note the reverse-polarity wiring of the diodes. This is necessary for proper matrixing of the signals so that one diode aids the passive resistive networks in the formation of $(L + R) + (L - R)$, and the other diode helps produce the $(L + R) - (L - R)$ signal. Thus, the respective left- and right-channel audio output is obtained. Capacitors C_1 and C_2 are filter units for the signals obtained by the demodulation (rectification) process. Resistors R_2 and R_3 form de-emphasis networks with capacitors C_3 and C_4.

The improved system for stereo FM demodulation is shown in Fig. 7-18, where the low-pass and bandpass filters are eliminated and the composite signal is applied to only two input sections. Again, the composite signal is obtained from a discriminator or ratio FM detector before de-emphasis and amplified. The 19-kHz amplifier uses resonant circuits; hence, only the 19-kHz signals are accepted, and others in the composite signal are rejected.

The 19-kHz pilot signal is applied to a doubler circuit consisting of full-wave diode rectifiers that produce successive alternations at a repetition rate twice that of the pilot carrier. These pulse a 38-kHz resonant circuit, and the flywheel effect produces a sinewave-type 38-kHz carrier for application to the demodulator, as shown in Fig. 7-19.

The composite signal is also applied to a 19-kHz bandstop filter, as shown in Fig. 7-18. This filter passes all signals except the 19-kHz pilot carrier; hence, its output contains the 50-Hz to 15-kHz demodulated $L + R$

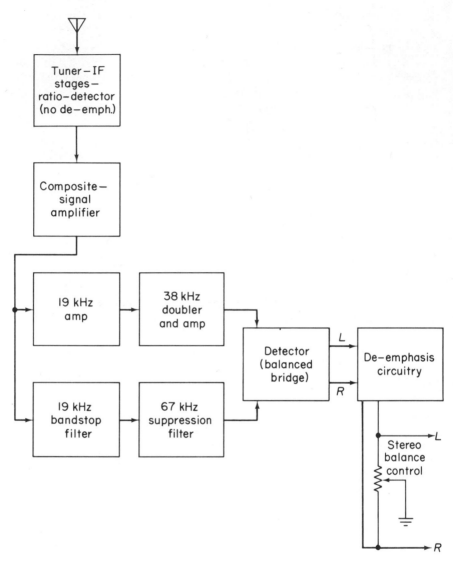

figure 7-18 multiplex FM decoder with balanced bridge demodulation

signals, the 23- to 53-kHz $L - R$ sideband signals, and 67-kHz SCA carrier signals, if these are being transmitted by the particular station being received. Next, these SCA signals are removed by the 67-kHz suppression filter. The remaining signals are then also applied to the balanced-bridge demodulation circuitry. The left- and right-channel signals are individually de-emphasized, as shown, and a common potentiometer across the two output lines serves as a stereo balance control.

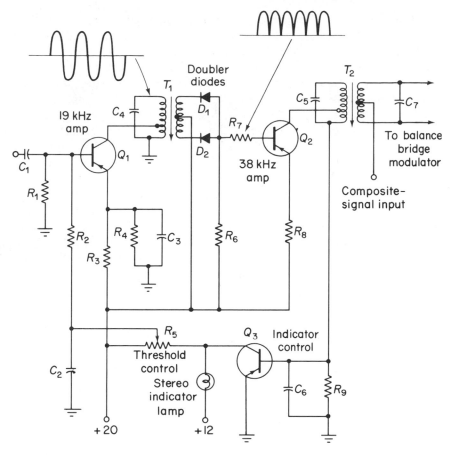

figure 7-19 pilot amplifier and doubler system

The intervening circuits between the 19-kHz amplifier and the demodulators are shown in Fig. 7-19. Transistor Q_1 is a straightforward amplifier, with transformer T_1 primary and the shunting C_4 forming the resonant circuit. Note that the 38-kHz amplifier Q_2 has a positive voltage applied to its base (via R_6 and R_7) from the $+20$-V source. Hence, Q_2 is reverse biased between base and emitter and does not conduct.

Note also that resistor R_9 is in series with the collector for Q_2, and if the latter does not conduct, no voltage drop occurs across R_9 and no forward bias is developed for Q_3. This transistor, an *npn*, requires a positive potential at the base with respect to the emitter. If Q_3 does not conduct, the stereo indicator lamp (in series with the collector feed line) has no current flow in it and does not light up, indicating reception of a mono signal or that the receiver is tuned between stations.

When a stereo signal is tuned in, the 19-kHz signal is applied to the input of Q_1 and amplified. The signal across the secondary of T_1 encounters diodes D_1 and D_2, making up a full-wave rectifier circuit. If, for instance, one alternation of the incoming signal is positive at the top of the secondary of T_1 (and consequently negative at the bottom of the secondary), diode D_2 conducts, rectifying that portion of the signal dropping across the lower half of the secondary between the center tap and D_2. Resistor R_6 completes a path back to the center tap, and the current flow through R_6 sets up a voltage drop across this resistor that is negative toward the D_2 and R_7 junction and positive at the other end. This negative potential now supplies the required forward bias for Q_2 and serves as a switching pulse to cause Q_2 to conduct.

Transistor Q_2, in turn, amplifies this pulse and applies it to the resonant circuit composed of C_5 and the primary inductance of transformer T_2. Since resonance is for 38 kHz, the pulse input produces a sinewave-type signal. Similarly, when the next alternation of the 19-kHz signal appears at the secondary of T_1, diode D_1 conducts and the current flow through R_6 is still in the same direction as that for D_2; hence, transistor Q_2 is again switched into conduction. Thus, successive full-wave alternations switch Q_2 into periodic conduction at a 38-kHz rate and form a sinewave carrier in T_2 for application to the bridge modulator.

During conduction of Q_2, collector current flow through R_9 causes a voltage drop that is positive at the top of this resistor at the base of Q_3. This positive potential supplies the required forward bias and Q_3 conducts, causing the stereo indicator lamp to glow. Resistor R_5 samples some of the potential in this circuit and regulates the forward bias on Q_1; hence, the gain. This adjusts for optimum performance for stereo reception, and by reducing the gain sufficiently, the system becomes insensitive to noise that might otherwise cause the indicator lamp to glow. Capacitors C_2 and C_6 filter signal components from these sections to obtain the necessary steady-state dc bias. Variable-core ferrite rods permit precise tuning adjustments of transformers T_1 and T_2.

The output system of Q_2, the 38-kHz amplifier shown in Fig. 7-19, is illustrated in Fig. 7-20. The composite signal mentioned earlier (minus the 19 and 67-kHz signals) is applied to the center tap of the secondary winding of T_2. The four diodes form a balanced-bridge system for processing the incoming 38-kHz signals in relation to the composite signals. Thus, as detailed for the system in Fig. 7-17, the 38-kHz subcarrier is reintroduced into the 23- to 53-kHz sideband structure and demodulated for the $L - R$ signal components. The composite signal applied at the center tap of the T_2 secondary appears (in phase) at any instant of time at the top and bottom of the bridge rectifier system. The incoming 38-kHz carrier, however, places out-of-phase signal across the top and bottom of the bridge for any instant of time, because when the signal potential is positive at the top of the secondary

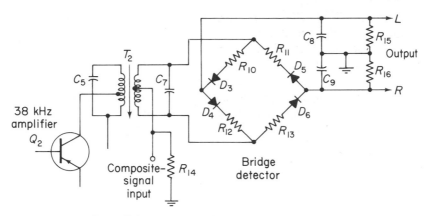

figure 7-20 balanced-bridge stereo demodulator

winding, the bottom of the winding is negative. Thus, phase relations occur between the signals in similar fashion to the balanced modulators and discriminator circuits discussed earlier. Capacitors C_8 and C_9 convert the rectified pulses to an average value voltage that varies in frequency and amplitude to conform to the audio-modulating component that had been contained on the original left and right channels.

From the rectification standpoint of the demodulation process, a positive alternation of the 38-kHz subcarrier signal across the secondary of T_2 would cause diodes D_3 and D_4 to conduct, establishing voltage drops across R_{10} and R_{12} within the bridge system, as well as across C_8 and R_{15}, with C_8 acting as the ripple-filter capacitor. For an incoming alternation of the 38-kHz signal, which causes the top of the secondary to be negative, diodes D_5 and D_6 conduct, producing an output across R_{16}. (Resistor R_{14} is the ground return to the center tap.)

For the composite signal applied at the center tap, assume initially that a positive alternation appears as against a negative potential at ground. In such an instance current would flow through both R_{15} and R_{16} simultaneously, with the path through diodes D_3 and D_6. For the next alternation (positive at ground and negative at the center tap), diodes D_4 and D_5 conduct, thus producing the required $(L + R) + (L - R)$ and $(L + R) - (L - R)$. During mono reception, only the demodulated $L + R$ audio components appear at the center tap of the T_2 secondary, and the positive and negative alternations produce identical voltage drops across resistors R_{15} and R_{16}. Thus, conduction of diodes D_3 and D_6 produces equal but opposite voltages across R_{15} and R_{16}. When diodes D_4 and D_5 conduct, the polarity of voltages across R_{15} and R_{16} reverses, but one is still opposite in polarity to the other. Thus, each output resistor develops the audio alternating-current signal for mono reception.

questions and problems

7-1. Explain briefly the essential differences between *direct* and *indirect* frequency modulation.

7-2. Since FM entails deviation of carrier frequency, what methods are used to stabilize the frequency of the carrier generator?

7-3. What are the purposes for using a *predistorter* in the indirect-frequency-modulation system?

7-4. What characteristics must a mixer circuit have to produce sum and difference signals for two signals applied to the inputs?

7-5. Give a brief description of the function of a *varactor* diode in frequency control.

7-6. How does a balanced modulator eliminate the carrier signal while still producing sideband-signal output?

7-7. Explain briefly how a discriminator produces a correction voltage for frequency-control purposes.

7-8. What precautions are taken to prevent the frequency-modulation components present in the phase discriminator from affecting reactance control of a carrier oscillator?

7-9. What is the average audio-frequency response capability that applies to FM stereo broadcasting?

7-10. In what manner is stereo transmission made compatible so that mono reception is not impaired?

7-11. Briefly explain the purposes for using a 19-kHz signal and a 38-kHz signal during stereo transmission.

7-12. What modulating signals in addition to the audio components are used for stereo transmission?

7-13. In what manner is the maximum permitted modulation for FM stereo broadcasting divided among the modulating signals?

7-14. If SCA transmission is also utilized, how is the modulation percentage then divided among all the signals involved in carrier modulation?

7-15. Briefly explain the purposes for using *pre-emphasis* and *de-emphasis* circuitry in FM transmissions, and indicate the *RC* used.

7-16. What are the essential circuit differences between the discriminator FM detector and the ratio detector?

7-17. What is the purpose for inductor L_4 in the circuits shown in Figs. 7-12 and 7-14?

7-18. What is the purpose for capacitor C_6 in Fig. 7-14?

7-19. In what menner is integration used to obtain audio output from a gated-beam detector?

7-20. What advantages, if any, does a gated-beam detector have over the ratio-detector method?

7-21. How are the 23- to 53-kHz sidebands obtained in an FM stereo receiver?

7-22. What circuit is used for combining the 38-kHz carrier with the $L - R$ sidebands, while also demodulating the composite thus obtained?

7-23. What method is used for deriving the 38-kHz signal from the 19-kHz signal obtained from the ratio detector of the receiver?

7-24. Explain briefly the manner in which the stereo signal lamp circuit senses the presence of a stereo signal.

7-25. In what manner are SCA signals minimized in FM receivers not designed to receive them?

7-26. Explain briefly in what manner the $L + R$ signals are processed through a receiver during reception of a mono signal.

7-27. In what section of stereo FM receivers does de-emphasis take place?

7-28. In what manner does a 19-kHz amplifier eliminate other components of the composite FM signal?

8

b/w television modulation
and detection

8-1. introduction

The basic principles of amplitude- and frequency-modulation systems covered in Chapters 6 and 7 (as well as the amplification circuitry covered in Chapters 3 and 4) apply also to television modulation and demodulation practices, both for black-and-white (B/W) and color. The sound portion of the television broadcast is frequency modulated, and this FM carrier is sent as an independent entity in addition to the television carrier. The latter, however, is amplitude modulated.

The camera tube uses an electron beam to scan an image focused on a photosensitive plate and acts as a transducer, just as a microphone does. Instead of converting sound in acoustical form to electric signals, the camera tube converts light and dark areas of a scene to changes in electric-signal amplitude. Since an abrupt change from light to dark produces sudden change in signal amplitude, high frequencies are involved. Thus, the frequency span of the video signals exceeds that of audio to a considerable extent and generates widely spaced sidebands during the modulation process.

To conserve spectrum space, part of the sidebands are suppressed, as described later in this chapter. Also, the frequency modulation uses a more narrow band than that for conventional FM broadcasting in the 88- to 108-MHz allocations. Despite such curtailment, however, the total bandwidth for a particular television broadcast is 6 MHz (compared to the approximate 10 kHz for standard AM and the maximum 200 kHz allocated to FM).

Such a 6-MHz (6000-kHz) frequency span requires special circuitry to obtain a wide bandpass in both the video IF amplifiers as well as the video amplifiers following the detector. In addition, various synchronizing and screen-blanking pulses must be sent from the transmitter so that the scanning sequences of the receiver are locked in with those at the transmitter. Thus, the video signals must also contain such pulses in precise timing and of proper duration to accomplish their purpose. Hence, the transmission and reception of television signals entails much more circuitry than found in radio broadcasting.

Tuners must also be of special design to handle both the VHF and UHF signals encountered in both black-and-white and color television. Since the detector system must derive the video signals plus the audio components, variations of the standard AM detector are necessary, as more fully discussed later.

8-2. TV transmission

A block diagram of the circuit structure of a complete television transmitter is shown in Fig. 8-1. The picture carrier generator is a crystal-controlled oscillator, and its signal frequency is multipled to obtain the desired VHF or UHF allocated frequency. The multiplier stages are usually followed by buffer Class C stages, which help to isolate the modulated final amplifier stage from the crystal oscillator and thus minimize loading effects on the oscillator. Even with crystal control, extreme changes in the load presented to the oscillator may affect frequency stability.

The final Class C amplifier is *amplitude modulated* by the composite video signal consisting of picture information, retrace blanking pulses, and synchronization timing pulses. The sound portion of television transmission consists of a frequency-modulated carrier that differs in frequency from the picture carrier and is independent of it. The FM transmitter portion contains the variable-frequency oscillator, reactance control circuitry, RF stages, and other circuitry described fully in Chapter 7.

A sideband filter is used for partial suppression of the lower-sideband signals of the video carrier to reduce the wide spectrum span that would otherwise prevail, as more fully described later. To use a single antenna to transmit the two carriers (AM video and FM sound) without interaction between the two, a diplexer network is used, consisting of a bridge circuit.

A master oscillator (31,500 Hz) is used for pulse-timing purposes and divided down to 15,750 Hz for the horizontal-sweep signal frequency and to 60 Hz for the vertical-sweep signal frequency. These signals are then used to form synchronization pulses, keying pulses for inserting or removing blanking, and other pulses as required. Thus, all blanking, sync, and keying

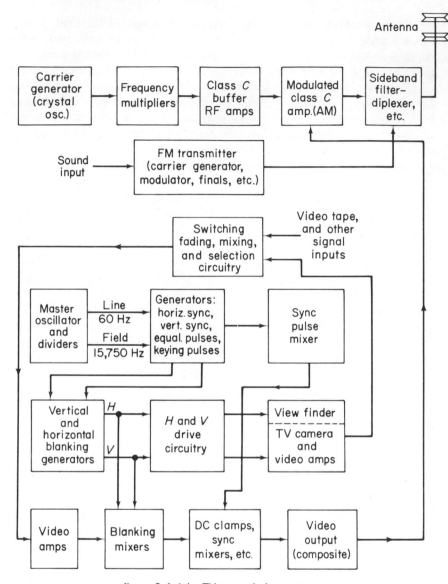

figure 8-1 b/w TV transmission system

signals are derived from a single master generator for the necessary synchronization linkage among them.

As shown, the television camera must have vertical and horizontal sync and blanking signals applied to it to form the composite video signal. The camera output is applied to mixer systems that are used to select specific cameras, video-tape signals, film, etc., as required. Additional mixing processes may be involved, as shown, and the composite video signal, which includes the picture signals, vertical and horizontal sync, blanking, etc., modulates the final Class C amplifier.

The sync and blanking generators are relaxation oscillators and were described fully in Chapter 3. Reference should also be made to Figs. 3-11, 3-13, and Fig. 3-15.

As mentioned in Section 8-1, sharp detail in a televised scene produces abrupt changes in video-signal amplitudes, thus involving high signal frequencies. Thus, for good reproduction at the receiver, a video-frequency span of from below 100 Hz to 4 MHz is highly desirable. For amplitude modulation, however, a modulating signal that extends to 4 MHz would mean that the sidebands above and below the carrier would produce an 8-MHz span for the picture signal alone. This would be the flat-band requirements, and resonant circuitry would introduce a taper that would extend the span to approximately 9 MHz. To reduce this prohibitively wide frequency span, some of the sideband information is eliminated. Since all the signal information is present in *either* the upper or lower sidebands, it was decided to suppress part of the lower picture sideband apan. For convenience in suppression circuitry (where it is difficult to have an abrupt cutoff) partial suppression is used, which results in a portion of the lower sidebands remaining intact (vestigial sidebands).

Thus, the total television span for a single station is as shown in Fig. 8-2a. Note that for the lower sideband section, modulation signals having frequencies higher than 1.25 MHz are entirely eliminated, and sideband signals with frequencies between 0.75 and 1.25 MHz are partially diminished, thus forming what is known as *vestigial sideband* transmission.

As shown, the sound carrier is allocated a spectrum space above the picture carrier and 0.25 MHz below the upper-end limit of the station's frequency span. Maximum deviation each side of the carrier's resting frequency is 25 kHz, and the highest modulating audio tone is 15,000 Hz. Thus, the deviation ratio [Eq. (2-8)] is

$$\frac{25 \text{ kHz}}{15 \text{ kHz}} = 1.67$$

With such a low deviation ratio, there are less than four significant sidebands present on each side of the carrier.

The total allocated bandwidth for each station is 6 MHz, as shown in Fig. 8-2a. The actual picture- and sound-carrier frequencies will, of course, vary for the individual channel allocations. For Channel 10 the individual carrier-signal frequencies are as shown in Fig. 8-2b. Note that the lower channel sound carrier is near Channel 10, while the upper picture carrier

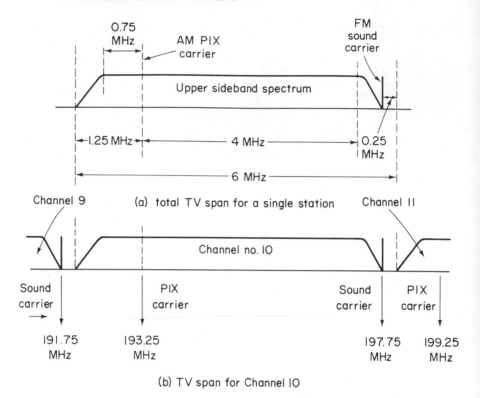

figure 8-2 TV station spectrum frequencies

(Channel 11) is also close to Channel 10. These close proximities of other carriers can cause some interference, and appropriate traps are included in receiver circuitry to minimize interference, as discussed later.

8-3.   scan and sync signals

A television camera tube (or a picture tube in a receiver) has an inductor network surrounding the tube neck for beam-deflection purposes. Both vertical-sweep and horizontal-sweep coils are included, and the combination is termed a *yoke*. The sweep oscillators generate the scan signals applied to the yoke, and during the time the horizontal oscillator is causing the electron beam to sweep across the tube face, the vertical oscillator gradually pulls the beam downward.

After the horizontal-line trace reaches the bottom, the vertical oscillator trips the beam back to the top and the scanning process is repeated. The white lines (against which the picture will be produced) are called a *raster*, which, as shown in Fig. 8-3, has an *aspect ratio* of 4 to 3. Thus, a raster could be 12 by 9 in. or 20 by 15 in., etc.

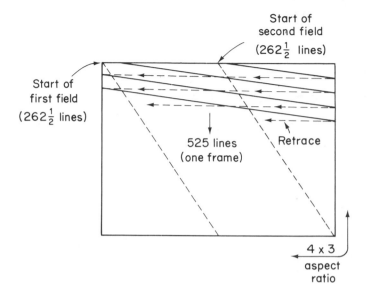

figure 8-3 interlaced picture-tube scanning

The standard scan rate is 525 horizontal lines (called a *frame*) repeated 30 times each second. This produces 15,750 horizontal lines per second (525 × 30) and, hence, comprises the frequency of the horizontal-sweep oscillator both at the station and in the receiver. The vertical oscillator's frequency is 60 Hz, and vertical blanking lasts about 850 μs (twice for each frame), with the retrace interval spanning about 40 horizontal lines. Hence, only about 480 horizontal lines actually are visible to convey picture information.

As shown in Fig. 8-3, the horizontal-line scan is not in sequential order, but instead the odd-numbered lines (1, 3, 7, etc.) are scanned first. Next the beam returns to the top and traces out the even-numbered lines, providing for a double scan of the scene being televised. This is known as *interlaced scanning* and reduces the flicker that would be evident otherwise. The process is similar to the practice of using a shutter in motion-picture projectors for blanking out the picture momentarily so that it can be projected twice for each frame.

When the screen is scanned once (262½ times) it is termed a *field*, with two such fields making up the complete frame (525 lines). As shown in Fig. 8-3, one field starts at the upper left and ends at the bottom center. The next field, however, starts at top center and ends at bottom right. This means that the sync timing for one filed differs from the next because of the one-half line starting and ending differences. Sync-signal timing relationships for two successive fields are shown in Fig. 8-4.

As shown, the horizontal sync signal is mounted on the blanking signal so that the retrace lines of the raster are eliminated during retrace. Picture information is present between the horizontal-blanking intervals.

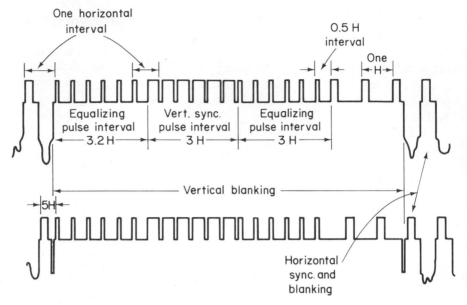

figure 8-4 sync for each vertical field

During the vertical sync interval, it is also necessary to blank out the retrace, though now for a longer period than was the case for the horizontal. During the lengthy vertical blanking, however, it is essential that the horizontal oscillators be kept in perfect synchronization with the master system; otherwise, it would be almost impossible to pick up synchronization immediately after the vertical interval has passed. Thus, to prevent loss of horizontal sync, a series of short-duration pulses (called equalizing pulses) are introduced at the start of the vertical blanking. These equalizing pulses are too short to trigger the vertical oscillator, but keep the horizontal oscillators in synchronization.

The vertical sync pulse is serrated to form a series of pulses having a short time duration between them. These, as shown later, are applied to an integrator system to form a signal having proper amplitude for triggering the vertical oscillator. The notches between pulses again maintain horizontal oscillator synchronization. Following the vertical sync pulse, additional equalizing pulses are included to maintain horizontal sync. The equalizing pulse rate is 31.5 kHz, and thus twice that of the horizontal scan of 15,750 Hz, so that the one-half line shift between fields will not leave the horizontal oscillator without sync pulses for alternate fields.

Horizontal blanking occurs in a time interval of from 10.16 to 11.4 μs. The horizontal pulse ranges from 5.18 to 5.68 μs. The vertical blanking period is from 833 to 1330 μs per field, with equalizing pulses 2.54 μs in duration.

The six vertical blocks making up the vertical sync-pulse interval have a duration of 190.5 μs. (For additional timing data, see Appendix E.)

The master oscillator and divider system shown in Fig. 8-1 is examined in greater detail in Fig. 8-5. The frequency of the 31.5-kHz master oscillator is maintained precisely by comparing it to a 60-Hz signal obtained from a precision-frequency ac line voltage, or from a crystal-controlled generator. A discriminator frequency control circuit, such as described in Chapter 7, utilizes a reactance circuit to make corrections for any variations between the two frequencies sampled.

A frequency-divider chain dividing in a 7, 5, 5, 3 sequence brings the 31.5-kHz frequency down to 60 Hz. This 60-Hz signal is applied to the discriminator control for comparison to the 60-Hz stabilized-frequency signal, as shown. The 60-Hz signal is also haped for balnking purposes, and equalizing pulses are gated to mount on top of the blanking at designated intervals. Horizontal sync and pedestal signals are shaped and mixed with

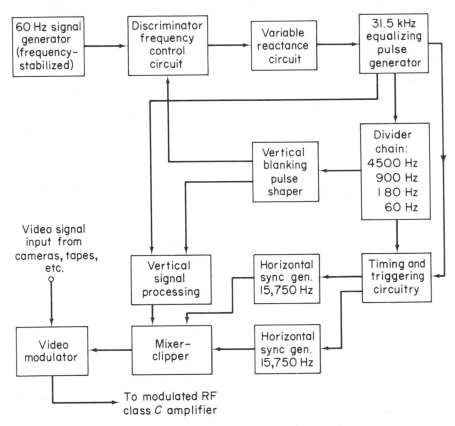

figure 8-5 composite video signal synchronization

the vertical signals and applied to the video modulator in conjunction with the video information.

While sync and shaping sequences may differ somewhat for the various transmitting systems, the basic operations are as shown in Fig. 8-5. These produce the various components of sync, blanking, equalizing, and video signals making up the composite video signal used to modulate the carrier amplifier. Delay systems may be used to shift pulses for mounting sync pulses on blanking pedestals. Specific circuitry produces pulses of desired duration and amplitude, and conventional mixing circuitry combines various signals as required.

8-4. monostable and delay circuitry

The circuit shown in Fig. 8-6 is a monostable type that produces an output pulse of a predetermined duration when triggered. This circuit is also known as a *start–stop* type or *single-shot* multivibrator. It differs from the conventional multivibrator, however, since it is not a free-running relaxation socillator (producing a continuous output signal regardless of input).

As shown, a positive-polarity input pulse is necessary for the circuit illustrated in Fig. 8-6. A negative potential (B_2) is applied to the base of the *npn* transistor Q_2 and furnishes a reverse bias between base and emitter, thus cutting off conduction in this transistor. During nonconduction of Q_2, no voltage drop occurs across the collector resistor R_4, and collector voltage has

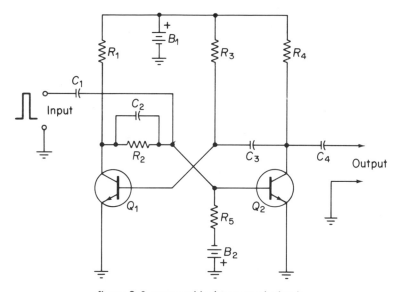

figure 8-6 monostable (start-stop) circuit

an amplitude equal to that of the source voltage B_1. Transistor Q_1 obtains the necessary forward (positive) bias applied to the base by R_3 and a negative potential for the emitter through the ground connection. Thus, Q_1 conducts fully and Q_2 is cut off.

Upon application of the positive-polarity trigger pulse to the input, the reverse-bias condition of Q_2 is overcome and, hence, Q_2 conducts. A voltage drop now occurs across resistor R_4 because of the current flow through it; hence, the collector potential is lowered. This decrease is felt at the base of Q_1 and reduces the foward bias and, hence, the conduction through this transistor. Now the voltage drop across R_1 decreases and the collector voltage for Q_1 rises. This increases the forward bias on Q_2, and the process is repeated, causing Q_1 to cut off current conduction and Q_2 to conduct fully. Thus, an output pulse is formed and its duration continues until the input trigger pulse initiates a reversal of the process.

When the input trigger pulse drops to zero, there is a time elapse before C_3 assumes its former charge level, which it acquired while Q_2 was conducting. As C_3 discharges, the positive potential applied to the base of Q_1 rises, providing the necessary forward bias to cause conduction. With Q_1 conducting and Q_2 again cut off by B_2, the original state has been reached and the output signal drops to zero. Thus, the output pulse can be regulated to have a duration as determined by circuit constants and not dependent on the duration of the input signal. The input signal (of short duration) is only instrumental in starting the production of an output pulse.

As shown in Fig. 8-7a, the single-shot multivibrator can be used to form a signal-delay system. A negative pulse of a predetermined duration is obtained from the one-shot circuit and differentiated as shown. The initial (negative-polarity) pulse obtained from differentiation is eliminated by a negative-signal clipper, leaving the second pulse [representative of time 2 (t_2) as compared to time 1 (t_1) for the input pulse to the one-shot multivibrator].

Delay systems can be formed by other circuitry, as illustrated in Fig. 8-7b and c. A pulse of a given duration (between time intervals t_1 and t_2) (Fig. 8-7a) is differentiated and applied to a relaxation oscillator that requires a negative pulse for synchronization. Hence, the output pulse will initiate at the t_2 time interval, representing a delay.

Another method for delay is to use a transmission line composed of series inductors and shunt capacitors (Fig. 8-7c). The unit can be made up of physical inductors and capacitors, or equivalent units formed by two lengths of wire, as more fully explained in Chapter 10. A signal applied to the input of such a line will reach the end of the line in a time interval determined by the number of inductors and capacitors and their respective values. Thus, if the line is tapped after the first inductor, as shown, a pulse will be obtained representing a later time interval (t_2) as compared to the original pulse entered

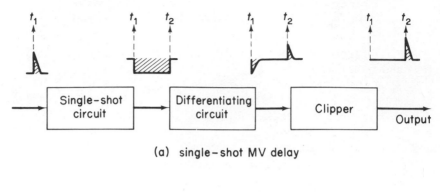

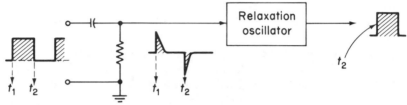

(a) single-shot MV delay

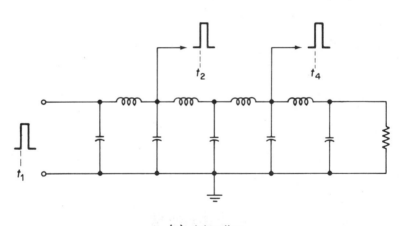

(b) relaxation oscillator delay

(c) delay line

figure 8-7 signal-delay systems

at t_1. Similarly, a time interval of t_4 (or higher) can be obtained as desired by tapping the line at a subsequent point.

The line is terminated by a shunt resistance equal to line impedance (again, see Chapter 10). The resistor absorbs signal energy reaching the end of the line and prevents reflections back along the line (which would disturb proper function).

8-5. equalizing-pulse insertion

There are many switching and mixing circuits suitable for inserting sync, equalizing, blanking, and vertical blocks into the video signal to form the composite picture used to modulate the TV carrier. Wide application is made of solid-state switching, gating, and mixing circuits, such as found in digital computers in which pulse gating is also a continual necessity. In such systems certain pulses of predetermined duration are used to key in (or key out) signals at the precise time required. As shown in Fig. 8-8, such gating or

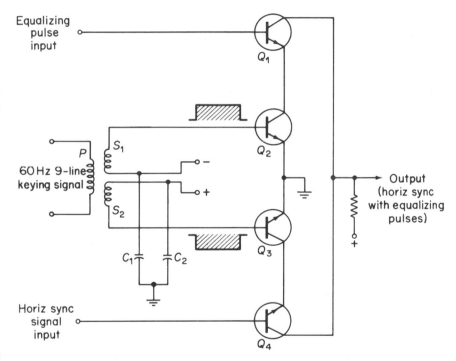

figure 8-8 gate circuitry for inserting equalizing pulses

switching circuits can also be used to insert signals, at the same time keeping out other signals during insertion.

The circuit shown in for the purpose of inserting 18 equalizing pulses during the vertical blanking time. During the time the equalizing pulses are inserted, the horizontal sync signals are keyed out. The keying signal consists of a pulse having a duration equal to nine horizontal lines and occurring at a 60-Hz rate. This keying pulse is applied out of phase to transistors Q_2 and Q_3, as shown. Since *npn* transistors are used in this instance, a positive signal at the base coincides with forward bias and affects conduction. A negative signal at the base reduces forward bias and decreases (or stops) conduction, depending on the magnitude.

Initially, however, assume that no keying pulse is present. The dc bias applied between the base and emitter of Q_2 is negative and, hence, holds this unit at cutoff. Since Q_2 is in series with Q_1, an open circuit exists for both transistors, and the steady stream of incoming equalizing pulses is kept from appearing across the output resistor. For Q_3, however, a positive bias is applied and, hence, conduction can occur. Thus, the incoming horizontal sync signals applied to the base of Q_4 are amplified and appear across the output resistor.

During the time the equalizing pulses are to be inserted into the composite signal being formed, the nine-line keying pulse appears at the base inputs of Q_2 and Q_3. For Q_2 the positive pulse overcomes the negative dc potential applied; hence, the resulting forward bias causes Q_2 to conduct. Now the equalizing pulses appear across the output resistor. For Q_3, however, the negative-polarity keying pulse applies reverse bias and cuts off this transistor. Consequently, the horizontal sync pulses are keyed out during the time the equalizing pulses are keyed in. Thus, the output train of pulses from this circuitry consists of timed horizontal sync pulses with a set of 18 equalizing pulses inserted at a time coincident with the vertical sync system.

Capacitors C_1 and C_2 place the bottom of windings S_1 and S_2 at signal ground, so the incoming signals appear between base and emitter. Transistors could also be used for phase inversion purposes instead of the transformer. Thus, a positive keying pulse could be applied directly to the base of Q_2 and inverted through a separate transistor for obtaining the necessary negative polarity for Q_3.

8-6.   vertical-block insertion

As shown in Fig. 8-4, there are six vertical blocks that follow the initial six equalizing pulses. To insert these, another gating system is employed, as shown in Fig. 8-9. Now the keying signal is of three-line duration. The three-line key is delayed by three lines with respect to the nine-line keying signal to make the block insertion after the initial six equalizing pulses.

As shown in Fig. 8-9, the horizontal sync with inserted equalizing pulses is applied to the base of Q_1 and appears across R_4 without influence by any gating circuitry. Transistors Q_2 and Q_3 are in series, with their combined output also appearing across R_4. Since Q_3 has a negative potential applied to the base (via R_3), the resulting reverse bias cuts off this transistor and prevents the entry of the vertical sync blocks appearing as a continuous train at the base of Q_2. When, however, the 60-Hz three-line keying signal appears at the base input to Q_3, the reverse bias is overcome by the positive potential of the keying signal, and Q_3 now conducts. Such conduction closes the series circuit with Q_2; hence, the equalizing pulses now appear across the output resistor R_4.

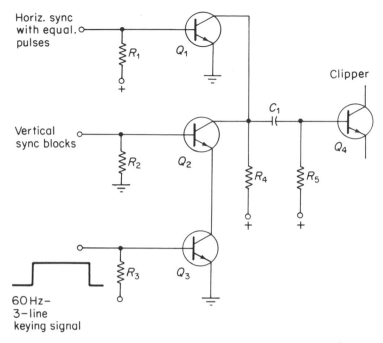

figure 8-9 vertical block insertion

The vertical blocks ride over the six center equalizing pulses. Consequently, this particular system has the advantage of straightening the leading edges of the vertical blocks because of the sharp rise time of the narrow equalizing pulses. Since the combining process is additive, a clipper follows the gating circuit to eliminate any overshoot that might occur. The transistors could, of course, be *pnp* types with appropriate reversal of bias polarities. Also, FET units could be substituted with proper voltages for gate, drain, and source elements.

8-7. *composite signal completion*

Component mixing of the vertical and horizontal sync signals, the vertical and horizontal blanking, and the video signals is shown in Fig. 8-10. The timed and composite vertical and horizontal sync signals are applied to the base of transistor Q_1 (a *pnp* type) and appear at the output without influence by the gating system. Since these signals have been timed with respect to blanking, they will combine with the latter across the output resistor.

Transistors Q_2, Q_3, and Q_4 are also *pnp* types and are biased for normal conduction. Thus, the appearance of the video signal at the base of Q_2 produces an amplified output across the common output load resistor. Each time the horizontal blanking signal appears at the base of Q_4, it opposes the

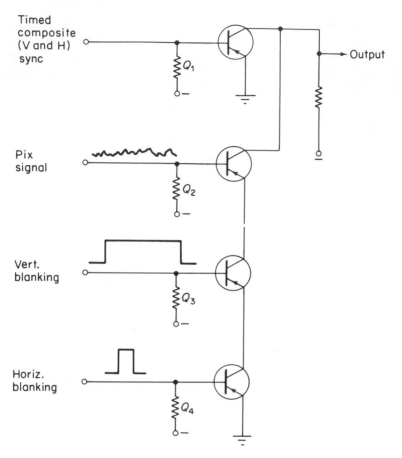

figure 8-10 component mixing of sync, blanking, and video

normal forward bias, and Q_4 is driven toward the cutoff point, eliminating the entry of the picture signal at Q_2 (which is undesired during blanking and sync intervals). The drop in conduction results in the appearance of the blanking pulse across the output load resistor. (It must be remembered that phase inversion occurs for signals between the base input and collector output).

Similarly, when the vertical blanking signal appears at the base input of transistor Q_3, the high positive potential nullifies the forward bias applied to the base resistor, and Q_3 is driven into the cutoff region. Again, the open circuit that occurs in the series chain of the three transistors prevents the video signal from appearing at the output. The vertical blanking signal is developed at the output, however, because of the sudden drop of voltage across the output resistor.

When either the horizontal or vertical blanking signals appear across the output resistor, the vertical and horizontal sync pulses are mounted on the blanking, as are the equalizing pulses and the vertical blocks. Between the blanking intervals appear the video signals, and thus the complete composite video signal has been formed and is ready for amplification to the level suitable for amplitude modulation of the TV carrier.

8-8. picture–sound diplexer

The diplexer shown in Fig. 8-11 permits the use of a single antenna system for both the picture-modulated carrier and the sound-modulated carrier. This system uses a balanced-bridge circuit to eliminate interaction between the video AM and the sound FM carriers. As shown, the output from the vestigial sideband filter (described fully in Chapter 10) is applied to the primary of a transformer arrangement consisting of L_3 and L_4. With a coaxial-cable feed to L_3 an unbalanced-line condition exists, because the inner conductor is above ground and the outer conductor (the shield) is grounded (see Chapter 10). The center-tapped L_4 converts the unbalanced input to a balanced line section and is referred to as a *balun*. As described in Chapter 10, such transformer and filter systems (at high frequencies) can be formed from coaxial-line sections.

As shown in Fig. 8-11, a basic four-arm bridge circuit is used, with two legs of the bridge consisting of coaxial-cable sections forming an inductance

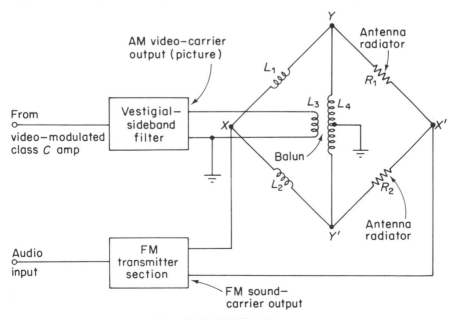

figure 8-11 TV-FM diplexer

and, hence, inductive reactance. These sections are designated as L_1 and L_2, while the other two legs of the bridge, R_1 and R_2, are the radiators of the transmitting antenna. These are represented as resistors, because propagated energy from an antenna is as though the energy were consumed by *radiation resistance* (see Chapter 12).

The FM sound carrier is impressed across points x and x', while the picture signal from the L_4 balun section appears at the bridge points y and y'. With a balanced bridge, equal voltages exist on both the reactive as well as the resistive legs. Hence, the sound-modulated carrier is present in both antenna radiators. Across points y and y', however, there is zero voltage for the sound-modulated carrier, since any potentials at hese points have like polarity, and no current flow occurs through L_4 from the sound-modulated carrier. Thus, interaction between the sound- and video-modulated carriers is eliminated.

Similarly, the video-modulated carrier applied to opposite points of the bridge at y and y' also sets up equal voltages across the legs of the bridge. As with the sound-modulated carrier, the video signal carrier will, therefore, be present in the antenna radiators. Because of the balanced-bridge system, no voltages for the picture carrier appear across points x and x' and again interaction is prevented.

8-9. b/w TV receiver

The basic sections of a black-and-white television receiver are shown in Fig. 8-12. Both the video- and sound-modulated carriers of the selected station are fed into either the VHF or UHF tuner section, as shown. The mixer stage heterodynes both the video- and sound-carrier signals with that produced by the local oscillator. The video IF amplifiers are tuned broadly for resonance to the intermediate-frequencies produced and reject the original carrier signals as well as the ones representing sum frequencies.

At the video detector, demodulation occurs to produce the picture signal. Also, additional heterodyning occurs and the sound and picture IF signals are mixed. Since the *difference* frequency between the sound and picture carriers is always 4.5 MHz, the sound IF circuits are tuned for resonance to this signal frequency. These, in turn, apply the signal to conventional FM-detector and audio-amplifier stages.

Signals from the video detector are applied to the AGC system for prevention of overload from local stations and to maintain a contrast level set by the viewer. The AGC is comparable to the AVC used for radios, and typical circuitry is discussed later.

The video signal with its sync and blanking signals is also applied to sync-separator circuitry, which derives the sync pulses from the composite

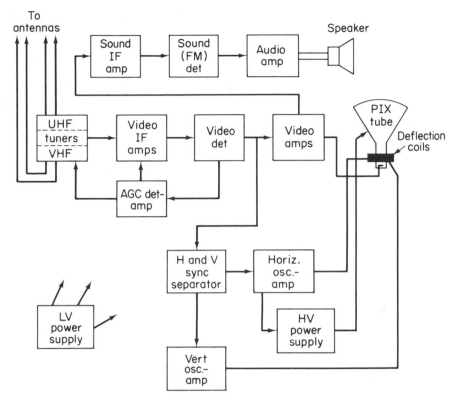

figure 8-12 b/w TV receiver sections

video signal. The sync pulses are applied to the sweep generators, one of 60 Hz for the vertical scan and the other 15,750 Hz for the horizontal sweep. Sawtooth sweep currents are developed by the sweep amplifier stages and applied to a double set of coils (called a *yoke*) surrounding the neck of the picture tube. The currents circulating within the horizontal and vertical coils control the deflection of the electron beam within the tube and, hence, trace out the raster on the picture-tube face.

A low-voltage power supply is present for filament and anode voltages for tube-type circuits, or for forward- and reverse-bias needs for transistor stages. In addition, a portion of the pulse waveforms developed in the horizontal output system is used to boost the low voltage supply and at the same time to develop a high voltage as required for the picture tube (over 10,000 V for black and white, and about 25,000 V for color tubes).

Some of the special circuitry associated with television receivers will be discussed in the remainder of this chapter. Many of the basic circuits employed in TV have, however, already been covered, and reference should be made to them for review purposes at this time. (See Sections 3-5 through 3-7, including Figs. 3-11 through 3-15. The initial portions of Chapter 4 should also be reviewed, particularly the references to RF stages. Section 5-7, should be referred to, especially Figs. 5-15 and 5-16.)

8-10. IF response curve

The IF response curve resulting from the mixing process in the tuner is shown in Fig. 8-13. Note that the progression of frequency points from left to right shows a rising frequency. Compare this with the station spectrum frequencies shown in Fig. 8-2. You will note that the curve in Fig. 8-13 shows

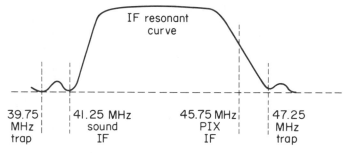

figure 8-13 frequency points on IF response curve

the picture IF above the sound IF for the station being received—in opposition to the incoming carrier frequencies shown in Fig. 8-2. This comes about because of the mixing process, when the frequencies shown in Fig. 8-2 are subtracted from that of the tuner oscillator.

As an example, assume that Channel 10, shown in Fig. 8-2b, is being received. The picture carrier of 193.25 MHz heterodynes with the local oscillator frequency of 239.00 MHz to produce an IF of 45.75 MHz. (Other oscillator frequencies could be used for this particular station, though the 45.75-MHz IF has been made standard for almost all modern receivers.)

The sound carrier for Channel 10 is 197.75 MHz, and when this mixes with the local oscillator signal frequency of 239 MHz, the result is a 41.25-MHz sound IF carrier. These two IF signals have the same frequencies for other stations, because the oscillator in the tuner changes frequency for other stations so as to obtain the same frequencies for the IF signals. Channel 3, for instance, has a picture carrier of 61.25 MHz, and the oscillator in the tuner would now be 107 MHz. When these two now heterodyne, an IF of 45.75 MHz is again obtained for the video IF.

The video IF of 45.75 MHz is set at approximately the 50 per cent point on the slope of the response curve. This decreases the abnormal gain that would otherwise result around this section because of the vestigial portion of the lower sideband. Since this remnant below the carrier would add amplitude equal to a similar portion above the carrier, the demodulated signals from the detector would be abnormal around this region and impair the desired flat response. With the video IF set down on the slope, as shown, the added amplitudes obtained from the vestigial portion of the carrier level off for the required flat response.

The dips in the response curve shown in Fig. 8-13 are caused by receiver traps designed to eliminate interference in the picture from adjacent channels (and from the receiver's own sound signal). Reference again to Fig. 8-2 will indicate that the lower adjacent channel (Channel 9 for the specific instance shown in Fig. 8-2b) has its sound carrier near the start of the Channel 10 video carrier. When this 191.75-MHz lower adjacent-channel sound carrier heterodynes with the 239-MHz oscillator signal, the result is a signal with a frequency of 47.25 MHz. Similarly, the upper-channel picture carrier of 199.25 MHz mixes with the local oscillator frequency of 239 MHz and an interference signal of 39.75 MHz is produced. These spurious signals would ride through the video stages and appear at the input of the picture tube where they would cause interference in the form of dark bars, herring-bone lines, and, on occasion, double images.

If the receiver has a good IF bandpass (3.5 to 4 MHz), three traps, such as shown in Fig. 8-14, are usually found. With receivers having a more narrow

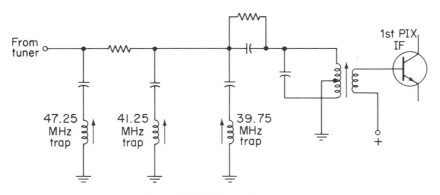

figure 8-14 TV-receiver traps

bandpass (around 3 MHz), several traps are not needed and often only a 47.25-MHz trap is used. The traps use a series capacitor and tuned inductance, and the series resonant circuit thus formed has a low impedance for the signal to which it is tuned, hence shunting such a signal.

8-11.   television detection

The basic demodulation principle of rectification discussed in Section 6-4 also applies to the detection of video signals, since they are also amplitude modulated. (Reference should also be made to Fig. 6-8 for a comparison of similarities to the picture detector for television.)

As shown in Fig. 8-15, the video and sound IF signals obtained from the

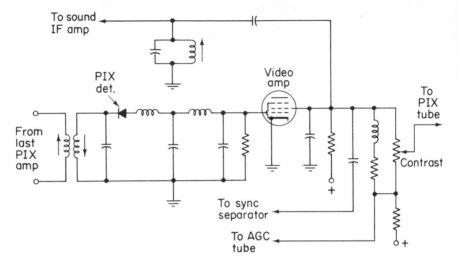

figure 8-15 b/w picture detector

last IF stage are applied to the picture-detector circuit. The arrows beside the transformer primary and secondary windings indicate metallic-core (slug) tuning sections, usually accessible at the top and bottom of the metallic-shield housing for the IF transformers. The inductors in series with the picture detector and the input to the video amplifier are *peaking* coils, described in Chapter 5. (Reference should be made to Section 5-7 and Figs. 5-15 and 5-16, where contrast and brilliancy control circuitry are shown, as well as input arrangements to the picture tube.)

The picture IF signal (usually 45.75 MHz) and the sound IF signal (41.25 MHz) are heterodyned in the video-detector circuit and produce a difference frequency signal of 4.5 MHz (standard for all receivers). This new IF sound carrier is then coupled to the sound IF amplifiers, either from the detector circuit or from the video amplifier stage, as shown in Fig. 8-15. Whether a tube or transistor video amplifier is used, some of the video signal is sent to the sync-separator circuitry and another portion is applied to the AGC system (as described more fully later).

The demodulated video signal obtained from the picture IF signal is shown in Fig. 8-16a. Note that the horizontal blanking level is at 75 per cent

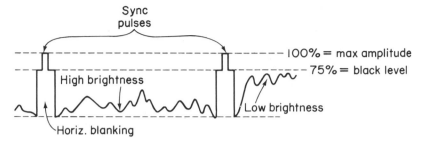

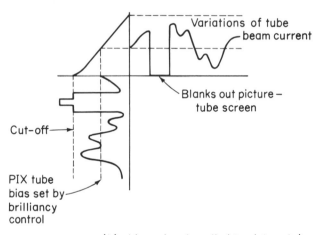

(b) video signal applied to picture tube

figure 8-16 demodulated video-signal characteristics

of the peak amplitude of 100 per cent reached by the horizontal sync pulses. (See also Fig. 8-4.) As the video signal increases in amplitude, it represents a *decreasing brightness level*, since it approaches the 75 per cent blanking level. Maximum brightness is represented by a low video-signal amplitude, as shown.

When the video signal is applied to the grid or cathode of the picture tube, the polarity must be such that the blanking levels cause a cutoff of the picture tube's electron beam, as shown in Fig. 8-16b. The brilliancy level most suitable to the viewer is set by the brilliancy control, which shifts the dc bias line of the picture tube. If the brilliancy is increased too much, it pulls the blanking below the cutoff point and, hence, retrace lines are no longer blanked out. Also, since the relationship of brilliancy to contrast is out of proportion, the picture becomes washed out in appearance.

The contrast control regulates the amplitude of the video signals and can be compared to the volume control in radios. Increased contrast darkens the picture background. This video-gain level is held at the predetermined position by the automatic-gain-control system described next.

8-12. automatic gain control (AGC)

Two basic automatic-gain-control systems are shown in Fig. 8-17, one using a solid-state diode (Fig. 8-17a) and the other a pentode tube plus a key-

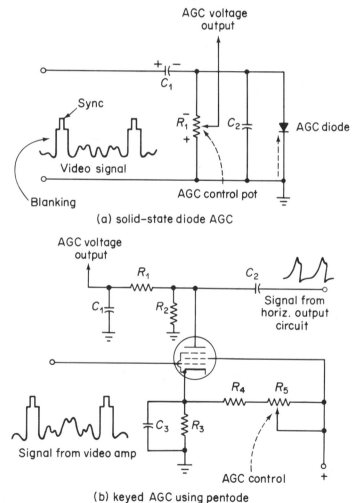

(a) solid–state diode AGC

(b) keyed AGC using pentode

figure 8-17 basic AGC circuits

ing feature (Fig. 8-17b). As shown in Fig. 8-17a, a separate diode is used for production of the AGC voltage, with the composite video signal applied at the input with a positve-going polarity. Electron flow is in the direction shown by the dashed-line arrow, and the diode conducts since the signal polarity is in the forward direction. During conduction, capacitor C_1 charges to the peak amplitude of the sync tips of the video signal. The time constant of R_1 and C_2 is made sufficiently long so that C_1 will maintain virtually a full charge during the intervals between sync tips.

The charge across C_1 also appears across the R_1–C_2 network, with a polarity as shown beside R_1. This voltage represents the AGC signal used to hold the bias level on the RF-IF amplifier stages at a constant level. Since R_1 is in the form of a potentiometer, it can be adjusted for the optimum degree of AGC voltage obtained for the RF–IF stages. Thus, although the *amplitude* of the picture signal changes constantly during reception, the sync-tip levels remain constant and, hence, hold the contrast level at that set by the contrast control.

Because of the long time constant, the AGC system of Fig. 8-17a is not influenced very much by rapid signal-amplitude changes unless they occur over a lengthy time interval. Thus, this particular method of AGC does not compensate for the rapid signal changes caused by video-signal reflections of passing airplanes. When a weaker station is tuned in, however, the modulated carrier has decreased amplitudes; hence, sync-tip levels are also lower. Consequently, less negative bias develops and the RF and IF stages amplify to a greater extent to compensate for the weaker station signal being received. For a strong local station, more bias is developed and the gain of the RF–IF stages decreases to prevent overloading.

Some of the disadvantages of the simple AGC system of Fig. 8-17a can be overcome by using the circuitry shown in Fig. 8-17b. This method is known as *keyed* AGC, and it provides for improved signal-to-noise ratios and a more rapid response to rapid changes in signal amplitudes. A triode tube or transistor (including FET units) can also be used instead of the pentode tube shown.

Note that the positive voltage applied to the screen grid of the tube is also diverted to the cathode by resistors R_5 and R_4. This positive potential appearing at the cathode makes the latter more positive than the grid, thus supplying sufficient bias to cause conduction cutoff. As with the circuit of Fig. 8-17a, a positive-polarity video signal is applied to the input of the keyed AGC system, as shown.

A positive-pulse waveform is coupled to the tube's anode using capacitor C_2 to isolate the anode's dc potentials. During the presence of this pulse, the plate is made more positive than the cathode and conduction could occur, except for the negative grid bias, which still holds the tube at cutoff. Thus, for conduction to occur the negative bias potential must be overcome.

In practice the circuit constants are selected so that only the sync tips appearing at the grid have sufficient amplitude to overcome the cutoff bias. Hence, neither the grid signal alone nor the anode signal alone cause conduction. Only when both signals are present does the tube conduct. Since the horizontal sync and blanking rate is 15,750 pulses per second, tube conduction occurs for every $1/15,750$ s.

During current flow, a negative AGC bias voltage develops across R_2, with pulse filtering performed by R_1 and C_1. For an increase in the incoming video signal, the grid is driven to a greater extent toward the positive region, hence, conduction *increases*. Consequently, a greater AGC bias is developed, and the gain of earlier stages is reduced to compensate for the carrier-amplitude increase.

Since this system permits conduction only during the presence of the sync signals (which are timed in unison with the horizontal keying pulse applied to the anode), there is not conduction between sync pulses. Thus, noise signals lying between sync tips have no effect on the AGC circuit. Also, since the time constant of the filter network R_1 and C_1 is for a ripple frequency of 15,750 Hz, the time constant is much shorter than that used for the circuit of Fig. 8-17a. Thus, this circuit is more sensitive to the fluctuating interference caused by rapid changes of carrier-signal levels and will develop a correcting AGC bias voltage more readily than the unkeyed type.

8-13. sync separation

The sync tips necessary for synchronization of the vertical and horizontal oscillators are extracted from the video signal and blanking levels, so no video-signal interference is present to upset good synchronization of the sweep circuitry. A single transistor (or tube) can be used, as shown in Fig. 8-18, and serves as both a sync separator and sync amplifier. On occasion a separate transistor may be used to provide for additional amplification when necessary.

As shown in Fig. 8-18, the sync-separator transistor has no forward bias applied, and thus is at or near the nonconduction state. On occasion some reverse bias may be applied so that only sync tips will have sufficient amplitude to cause conduction and, hence, provide for an output, depending on transistor parameters. If an *npn* type is used, as shown, the incoming signal must have a positive polarity so that the sync tips provide sufficient forward bias (making the base positive with respect to the emitter) to permit conduction. For a *pnp* type, of course, the input signal would have a negative polarity.

The sync pulses are applied to respective processing networks required for synchronization of the vertical and horizontal relaxation oscillators. As shown in Fig. 8-4, both the equalizing pulses and vertical blocks have the

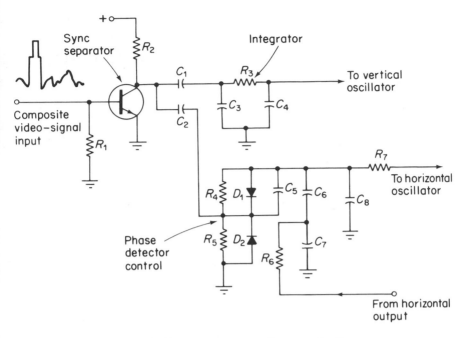

figure 8-18 sync separator and output circuitry

same repetition rate, but their durations differ. Thus, while each type pulse will synchronize the horizontal oscillator, the short duration of the equalizing pulses prevents triggering of the vertical sweep. The longer-duration vertical blocks are used to trigger the vertical-sweep oscillator by applying them to an integrator circuit, which permits a voltage buildup for successive blocks.

In Fig. 8-18 the integrator consists of R_3, C_3, and C_4. It has a long time constant for the vertical-block pulses; hence, the short-duration equalizing pulses occur in too short a time to permit a voltage buildup, as shown in Fig. 8-19. The capacitors charge slightly for each equalizing pulse, but the interval between pulses is longer than the pulse duration; hence, the capacitors discharge. When the vertical blocks arrive, their longer duration places a greater charge across the capacitors. Since the time interval between them is smaller than their duration, there is not sufficient time to discharge much energy. Thus, as each block appears the voltage builds up until it has sufficient amplitude to trip the vertical relaxation sweep oscillator.

After the vertical system has been triggered, the arrival of the last six equalizing pulses permits the discharge of the integrator capacitors, as shown. The discharging of the two capacitors places the circuit in readiness for the next vertical trigger and retrace one sixtieth of a second later. Thus, horizontal sync and equalizing pulses do not trigger the vertical oscillator because of the step-filter function of the integrator circuitry.

As also shown in Fig. 8-18, pulses are applied to a phase-detector control

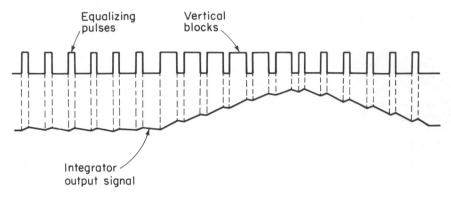

figure 8-19 formation of vertical trigger

system for synchronization of the horizontal sweep oscillator (also a relaxation type). The sync pulses are applied to the junction of the phase-detector diodes D_1 and D_2, as shown, permitting both diodes to conduct. A signal from the horizontal output amplifier circuitry is applied to the junction of capacitors C_6 and C_7, as shown. Essentially, a bridge is formed by the diodes and capacitors, and with in-phase voltages appearing at the diode junction and capacitor junction, a zero voltage appears across the top and bottom (ground) of the circuitry. In such an instance the horizontal oscillator is synchronized and generates the proper frequency.

If the horizontal oscillator drifts from the desired frequency, the phase of the signal fed back to the bridge phase detector no longer coincides with that of the incoming sync pulses, and the bridge becomes unbalanced. Consequently, a voltage now appears at the output (across capacitor C_8 and at R_7), which is applied to the input of the horizontal oscillator for frequency-drift correction.

8-14. sweep output system

Typical solid-state vertical- and horizontal-sweep systems are shown in Fig. 8-20. For the vertical output shown, no coupling transformer is used between the output transistor and the deflection coils, as is the case with tube-type output amplifiers. The low-impedance sweep coils are matched to the low impedance of the transistor output. Occasionally, when more precise matching is required, a transformer may be found in solid-state vertical-sweep systems.

The horizontal output system is always more complex than the vertical, since the high-voltage generation is incorporated in the circuitry, plus pulse takeoff for application to the horizontal phase detector, the AGC system, etc. A horizontal output pulse is developed that is stepped up to the required

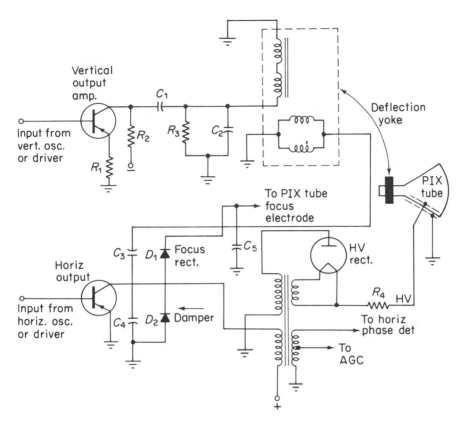

figure 8-20 sweep output circuitry

amplitude and rectified, as shown. Depending on picture-tube size, voltage may range from approximately 6000 to over 15,000 V for larger-screen tubes. (Color tubes, as discussed in Chapter 9, require 25,000 V or more.)

For the circuitry in Fig. 8-20, the sweep signals are coupled to the horizontal-sweep coils by capacitor C_3. The sawtooth signal appearing here is also rectified (and filtered by capacitor C_5) for the focus electrode of the picture tube. A damper diode eliminates transient voltages that result from the collapsing fields of the horizontal deflection coils at the declining portion of the sawtooth sweep signal. Since the damping diode functions as a rectifier, the resultant unidirectional current flow thus developed is often used to boost the low-voltage amplitudes, and thus to reduce the requirements of the low-voltage power supply. As shown for the color-sweep system in Chapter 9, a double boost (boosted boost) is also possible from damper voltages. Voltage boost may range anywhere from 50 to well over 200 V, as required.

The high voltage is applied to the inner conductive coating of the picture tube. This conductive coating on the inside of the bell of the tube consists of a collodial graphite and is sometimes referred to as an *aquadag* coating and second anode. It has the dual function of increasing electron-beam velocity within the tube and collecting the secondary-emission electrons released when the high-velocity beam scans the phosphor material coating on the inside of the picture-tube face. With many picture tubes, an additional conductive coating is applied to the outside of the glass so that the inner and outer coatings (with the glass between) perform as a capacitor and thus filter the ripple component of the rectified energy.

questions and problems

8-1. To what degree are sidebands suppressed in television broadcasting, and for what purpose?

8-2. What bandwidth does the picture signal occupy and how does this compare to the bandwidth used for sound transmission of the TV station?

8-3. Reproduce Fig. 8-2b, but show the signal frequencies involved when Channel 12 is the one to which the set is tuned. (See Appendix C.)

8-4. Define the terms *raster*, *aspect ratio*, *field*, and *frame*.

8-5. Briefly explain in what manner interlaced scanning is utilized in television broadcasting and reception.

8-6. Briefly explain the purpose for equalizing pulses and vertical blocks.

8-7. What are the specific signal frequencies utilized in television (obtained from a master oscillator having a signal frequency of 31.5 kHz)?

8-8. Briefly explain the features of a single-shot multivibrator that make it useful in communication systems.

8-9. Briefly explain how equalizing pulses are inserted by precision gating into the composite video signal.

8-10. What is the advantage of placing generated vertical blocks over equalizing pulses?

8-11. Briefly explain the purpose and function of the diplexer used in TV broadcasting.

8-12. Why is the final sound IF always 4.5 MHz in TV? Explain briefly.

8-13. Why is the video IF carrier signal set at approximately the 50 per cent point of the slope of the response curve?

8-14. What is the purpose for designing television-receiver IF stages to produce a curve having dips at certain frequencies, as shown in Fig. 8-13?

8-15. What circuitry functions are performed by the brilliancy and contrast controls when adjusted to increase or decrease brilliancy and contrast?

8-16. What is the purpose for an AGC system, and in what manner is the keyed type superior to the basic AGC circuitry?

8-17. Explain the purpose for the integrator circuit that follows the sync separator.

8-18. What is the purpose for a phase detector prior to the horizontal oscillator circuitry?

8-19. Briefly define the terms *yoke*, *aquadag*, and *secondary-emission electrons.*

8-20. What is the primary purpose for a damper diode in the horizontal-sweep output systems of TV receivers?

8-21. In what manner is a secondary feature of the damper utilized to boost the low-voltage-supply amplitude?

8-22. What commonly consitutes the filter capacitor for the high-voltage system of a TV receiver?

8-23. What is the function of the high voltage applied to the second anode of the picture tube?

9
color TV modulation and detection

9-1. introduction

The present-day color television system was authorized by the FCC in 1953 at the recommendation of a study committee of industrial engineers (National Television Systems Committee). This NTSC color system superseded an early color telecasting method of the Columbia Broadcasting System, which had also been approved by the FCC several years earlier. The CBS system, however, lacked compatibility with the 60-Hz vertical sweep and 15,750-Hz horizontal sweep of the conventional system used in the United States and could not be received on ordinary black-and-white receivers unless the vertical and horizontal sweep rates were capable of being switched.

The CBS system also gave poorer definition because of the reduced line-per-frame rate of 405 instead of the current 525. (The horizontal rate was 29,160 Hz instead of 15,750 Hz and, hence, 202.5 lines per field instead of 262.5.) A color wheel was used at both the transmitter and receiver and rotated before the camera tube (or receiver picture tube). The wheel had six filter segments of red, blue, and green (two of each), which rotated 1440 revolutions per minute (rpm). This system uses field-sequential scanning; that is, the first field made up of odd-numbered lines is scanned in red, while the second field (even-numbered lines) is scanned in blue. The first field of the next frame (again odd-numbered lines) is scanned in green, while the second field of the second frame is scanned in red. This sequence continued with a standard two-to-one interlaced scan. Thus, for six fields, each line will have been scanned once for each color.

The CBS system, while not used in standard broadcasts for public use, is still employed in some commercial applications because of the simplicity of the color pickup system in contrast to the three camera tubes needed in the NTSC system, with its accompanying need for optical-path systems, care in color-registration techniques, and general circuitry complexity. The field-sequential color system was used in some of the Apollo moon flights, with a 15-lb color camera in flights 10 and 11 using a color wheel spinning at 600 rpm with six color sections, two of red, two of blue, and two of green. Conversion equipment was used on earth to provide a 30-frame-per-second rate. Similarly, such field-sequential systems are occasionally used commercially for color pickup and converted to the compatible NTSC system and then telecast.

For our present-day NTSC system, considerable engineering design and development were necessary to include the color signals within our present 6-MHz bandwidth and still provide compatibility so that color telecasts could be received in black and white on B/W receivers as well as in color on color sets. As mentioned in Chapter 8, the many sidebands necessary for fine-detail video already entailed the suppression of some of the lower sideband spectrum. Thus, it was now necessary to include additional signals within the standard black-and-white transmission to convert it to a color telecast without undue cross interference between the black-and-white carrier and sidebands and the newly introduced color signal information. Hence, the color receiver must not only include all circuits normally used for black-and-white reception, but a number of others for handling and processing the color-data signals. These circuits and systems are covered in this chapter, both for the transmission and reception areas.

Since the factors of bandwidth, separation between the sound carrier and the video carrier, and the basic principles of the scanning processes applying to black-and-white transmitters and receivers also hold for color, these factors should be reviewed and understood before a study is undertaken of the complex color transmission–reception system. In color television not only are various new circuits required, but the vertical and horizontal scanning rate must be altered for proper integration of color vs. black-and-white signals. Despite such a sweep-rate change, however, compatibility is still achieved, as described herein.

9-2. color TV system

In the amplitude-modulation process of an RF carrier by a composite video signal (containing blanking pulses, sync, picture signals, etc.) the sideband signals produced tend to cluster around the harmonics of the horizontal sweep-frequency signal. Hence, groups of such bunched sidebands occur at specific spectrum intervals from the video carrier. This tendency for the sidebands to shift slightly toward the harmonics of the sweep signal is fortunate, since it permits the insertion of the color signals into the gaps between such clusters of sidebands.

For this reason the frequency of the color signals must be produced by a carrier that will assure their exact placement within the existing black-and-white video spectrum. Also, as with multiplex FM discussed in Chapter 7, the additional carrier cannot be accommodated within the existing frequency span; hence, it must be suppressed at the transmitter. Consequently, the subcarrier originally used to produce the color sidebands must be reintroduced in the receiver for proper demodulation.

A suitable subcarrier frequency would be an odd multiple of one half the horizontal-sweep signal frequency to assure proper sideband placement. A subcarrier having a frequency that is too high would result in crowding too near the 4-MHz span from the main carrier. The result would be a restricted color bandwidth and loss of fine detail. If the subcarrier frequency is made too low, interference might result from a heterodyne action between the main picture carrier and the color-signal sidebands.

A compromise between the undesirable frequency limits of the subcarrier is reached when one half the horizontal signal sweep frequency is multiplied by 455. Unfortunately, however, if the standard 15,750-Hz black-and-white horizontal-sweep frequency is used for this purpose, the desired odd-harmonic relationship could not be obtained with respect to the sound carrier accompanying the video signal. Hence, the horizontal-sweep frequency for color receivers is 15,734.264 Hz, which results in a subcarrier frequency of 3.579545 MHz (3.58 MHz) (multiplication by 455 of one half the new 15,734.264-Hz frequency).

As explained in Chapter 8, when the horizontal-sweep frequency is multiplied by 2 we get the master-oscillator frequency. When this is divided down by 7, 5, 5, and 3, we obtain the vertical-sweep frequency (60 Hz for black-and-white). For color transmission, the 15,734.264 Hz produces a vertical scan frequency of 59.94 Hz. Both the 59.94- and 15,734.264-Hz signals are sufficiently close to the normal frequencies generated by the vertical- and horizontal-sweep oscillators in the receiver to permit synchronization. Thus, lock-in is obtained without need for sweep-frequency adjustment when changing from black-and-white to color reception.

For color transmission, the other factors relating to black-and-white television still apply. Thus, the aspect ratio is still 4 to 3; the picture carrier is still 1.25 MHz above the lower end of the channel; the scan is still interlaced 2 to 1; and the sound carrier is still 4.5 MHz above the picture-carrier frequency with a permissible deviation of 25 kHz each side of center.

As shown in Fig. 9-1, the color camera contains three separate pickup tubes with the optical system indicated. This filters (from the image picked up

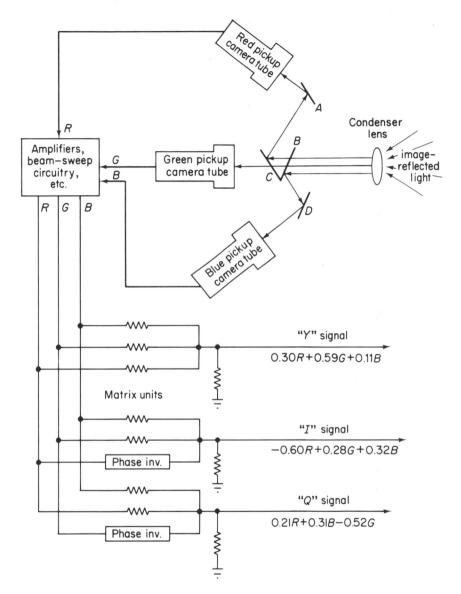

figure 9-1 color-camera optical system

by the lens system) the three primary colors of red, blue, and green (the *additive* color principle). The filters consist of dichroic mirrors indicated as B and C in Fig. 9-1. These have the characteristic of reflecting light of only one color and are used with front-surface mirrors A and D. Thus, when a televised scene has various colors, the dichroic mirrors separate the individual red, blue, and green components from the scene and channel these signals to the input section of the proper camera tube. Thus, red images are reflected by dichroic mirror C, and directed by front-surface mirror A to the upper camera tube, as shown. Blue images are reflected by mirror B, with front-surface mirror D reflecting the image to the lower camera. Green images ride through both C and D mirrors and are impressed on the center tube, as indicated.

The matrix section shown in Fig. 9-1 has two purposes, one of which is to obtain a luminance signal, which corresponds to the black-and-white signal of monochrome transmission, and the other to combine the three primary color signals of red, blue, and green into two basic signals to conserve space in the channel spectrum.

The luminance signal is often designated as the Y signal, and the proportions of red, blue, and green combined by the matrix resistors are as follows: 0.30 for red, 0.59 for green, and 0.11 for blue. These proportions are necessary to compensate for the manner in which color densities are perceived by the human eye. If the signals sampled were of equal amplitude for red, blue, and green, the eye would not respond evenly and some colors would appear to have greater intensity than they actually possess.

The two basic color signals produced by the two lower sections of the matrix unit consist of an I (in-phase) signal and a Q (quadrature) signal, which are essentially red signals minus the luminance (Y) component and the blue signal minus the Y. The I signal has the following proportions: -0.60 of red, 0.28 of green, and 0.32 of blue (-0.27 of blue minus Y, and 0.74 of red minus Y). The Q signal is 0.21 of red, 0.31 of blue, and -0.52 of green (0.41 of blue minus Y, and 0.48 of red minus Y).

The color circle shown in Fig. 9-2 illustrates the phase relationships for the various colors, with the primary ones of red, blue, and green indicated. Note the $B - Y$ signals are displaced 33° clockwise from the Q signals, while the $R - Y$ signals are also displaced 33°, but from the I signals instead of the Q. As shown in Fig. 9-1, a phase inverter is used to change direction by 180° to obtain the minus component.

While various circuit combinations can be used for modulation and sub-carrier suppression, the system shown in Fig. 9-3 is typical of the essentials for obtaining the modulating signals. As with the standard black-and-white systems described in Chapter 8, conventional circuitry generates the carrier and multiplies the frequency to that required. Additional Class C amplifier stages raise power to the level needed, and amplitude modulation is used for the picture signals, as with black-and-white transmission.

In addition to the crystal oscillator generating the primary video carrier signal, another oscillator is employed for generating the 3.579545-MHz subcarrier signal, which is suppressed in the balanced modulators after

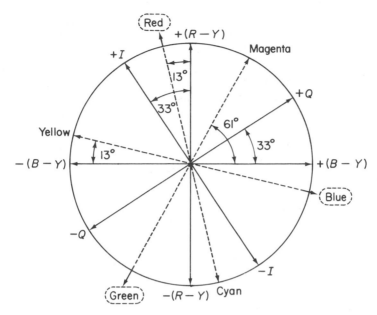

figure 9-2 color TV phase relationships

the modulation process has produced the necessary sidebands. (See Section 7-6 and Fig. 7-7.)

As also shown in Fig. 9-3, the three signals (Y, I, and Q) obtained from the matrix system shown earlier are applied to low-pass filter networks for setting the desired bandwidth. To minimize cross talk during color reception, the I signal is transmitted with one sideband extending to 1.48 MHz, with a vestigal sideband section approximately 500 kHz wide. The Q signal is transmitted as a double-sideband signal, each of which is 500 kHz distant from the frequency used for the subcarrier (but which is suppressed at the transmitter). The chroma, plus total bandpass, is shown in Fig. 9-4.

The I and Q signals are fed to the balanced modulators, as shown in Fig. 9-3, where they modulate two subcarriers, each of 3.579545 MHz but separated by a 90° phase difference with the I leading the Q. In the balanced modulators, the subcarrier signals, plus the I and Q, are suppressed and the only output consists of the sideband components.

A burst gate synchronized by the 3.58-MHz oscillator produces a minimum of eight cycles of the 3.58-MHz signal, which is mounted on the horizontal blanking level following the sync pulse (known as the back porch), as shown in Fig. 9-5. This burst signal is used in the receiver for synchronization of the 3.58-MHz oscillator, which replaces the missing subcarrier, as described later.

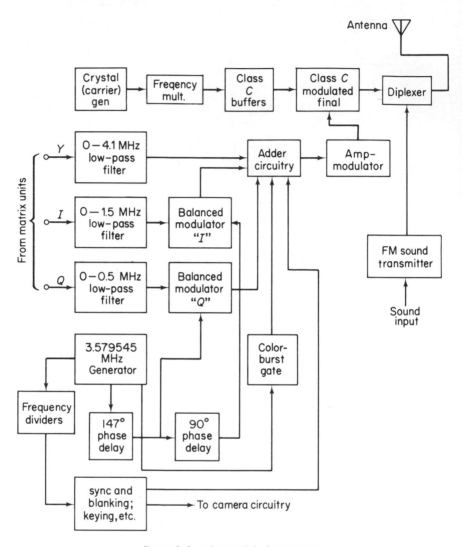

figure 9-3 color modulation system

The phase of the burst signal is 57° ahead of the *I* signal, as shown in Fig. 9-3, with the *I* leading the *Q* by 90°. All signals are combined in the adder circuitry, including the normal vertical and horizontal sync and blanking. Additional amplification may be employed to bring the composite color signal to the level required for amplitude modulation of the carrier.

When filter circuits are used to obtain specific bandwidths, the narrowing of the bandpass introduces signal delay. Thus, the low-pass filters shown in Fig. 9-3 upset the time coincidence of the *Y*, *I*, and *Q* signals. Thus, delay-

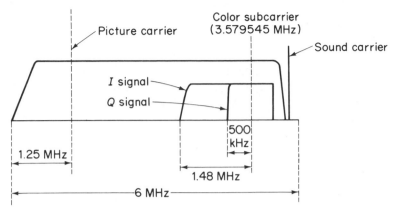

figure 9-4 chroma channel bandpass

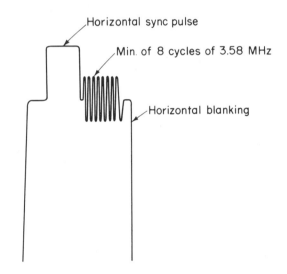

figure 9-5 subcarrier burst on blanking

producing sections of transmission lines are used to cause the I signals to have the same delay as found in the Q circuitry, because of the narrower bandpass, with a delay also included in the Y circuitry to compensate for that encountered in the I and Q. Delay compensation must also be introduced into the receiver, as described later. (See also Chapter 10.)

The subcarrier burst shown in Fig. 9-5 is not transmitted during the vertical pulse interval nor during black-and-white transmission. The burst has no effect on reception of a color signal by a black-and-white receiver, nor does the slight difference in the vertical- and horizontal-sweep frequencies used in color transmission.

9-3.   color receiver

The basic circuits included in a color television receiver are shown in Fig. 9-6. The luminance (Y) signals that constitute the black-and-white transmission for compatibility are applied to the cathodes of a three-gun color tube. The inner faceplate of the tube contains a series of red, blue, and green phosphor dots, which must be struck by the respective electron beams from the cathodes. Thus, for black-and-white reception all three phosphors must fluoresce, and the blend of the three colors produces black and white. Hence, if one gun becomes inoperative in such a picture tube, black-and-white reception is no longer possible.

Tuner and video IF stages in the color receiver compare to those in black and white except that precautions must be taken to assure a wide bandpass in the IF stages for good color detail (about 4.2 MHz). The wide bandpass and greater tendency for interference require a greater degree of sound trapping. Hence, the sound carrier is weaker at the video detector, and a separate sound detector is used. The final sound IF is 4.5 MHz, the same as in the black-and-white receiver.

The chroma (color) signals are obtained from the video amplifier and applied to the demodulators through the bandpass amplifier, as shown in Fig. 9-6. The I and Q signals (in relation to Y) have minus quantities of green; hence, this color must be restored. Matrix units (resistive networks) mix the Q and I signals to proper proportions, and after demodulation the chroma signals are $G - Y$, $B - Y$, and $R - Y$.

The three chroma (minus Y) signals are applied to the individual control grids of the three-gun tube, as shown. During reception of a black-and-white picture on a color receiver, interference can result if the video signals pass through the bandpass and color-demodulator circuits. To prevent this, a color-killer circuit is used to disable the bandpass amplifier and prevent any signal entry. (The function of these circuits will be described later.)

The 3.58-MHz burst signal on the pedestal of the horizontal blanking is increased in amplitude in the burst amplifier and applied to a phase detector. A crystal-controlled 3.58-MHz oscillator generates the required subcarrier signal, which is also applied to the phase detector and compared with that received from the burst amplifier. If any phase difference exists, the reactance-control circuitry produces a correction voltage that is applied to the crystal oscillator to compensate for any drift. The process is similar to that described in Chapter 7 for the discriminator control of an FM carrier, and illustrated in Fig. 7-8.

Conventional sync separation and sweep circuitry are used, as in black-and-white receivers. Convergence circuitry is included for precise adjustment of beam alignment within the picture tube. Each beam must hit its respective

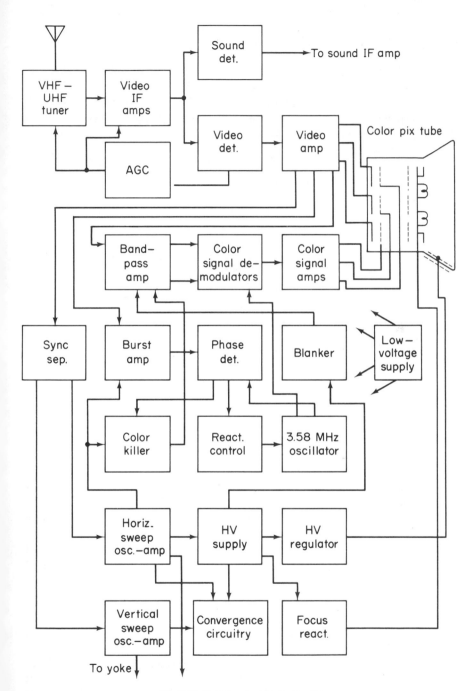

figure 9-6 color-receiver circuits

dots precisely or false colors will appear and may be visible as rims contaminating sharp changes in detail in the color scenes.

9-4.　*color detector and video amp*

A typical demodulating system for color television is shown in Fig. 9-7a. Signals from the collector of the last picture IF amplifier are applied to the sound detector, where the picture and sound IF signals are heterodyned

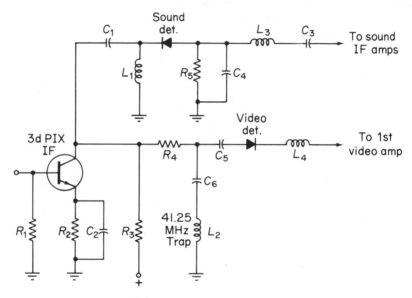

(a) demodulator for color TV

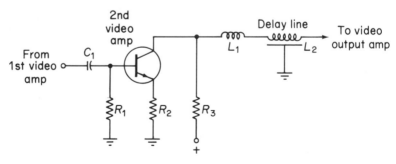

(b) basic color video amplifier

figure 9-7　color detector and amplifier circuits

to produce the 4.5-MHz IF signal in similar fashion to the process described for the demodulator in black-and-white receivers.

Signals at the collector are also applied to the video diode detector, as shown, and here a 41.25-MHz sound trap is included to minimize interference at the picture tube. Coupling capacitor C_5 blocks the dc voltage applied to the collector, and L_4 is the conventional series peaking coil, as also used in black-and-white receivers.

In solid-state receivers, two video amplifier stages may be present to bring the video signals to the level required for application to the picture-tube input. A basic video amplifier is shown in Fig. 9-7b, with L_1 again representing a peaking coil. Since the luminance (Y) signal has a wider bandwidth than the chroma channels, a delay must be introduced to achieve the same delay encountered by the color signals in the narrower bandpass. By providing time coincidence for the various signals, a sharper picture is obtained (blurring eliminated).

9-5.   bandpass and blanker circuitry

A tube-type bandpass amplifier and blanking system are shown in Fig. 9-8. The bandpass amplifier has the dual function of increasing chroma signal levels to that needed for proper demodulation and preventing sync and

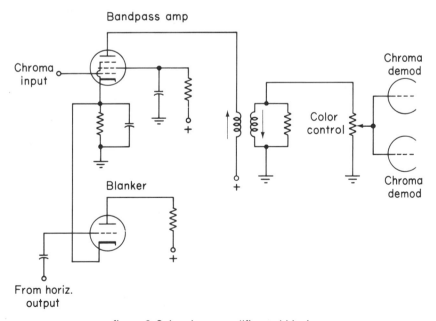

figure 9-8 bandpass amplifier and blanker

burst signals from appearing at the demodulation section. Undesired signals are held back by gating the bandpass tube into periodic nonconduction by a special gating circuit called a *blanker*. A gate pulse is obtained from the output section of the horizontal-sweep system and used to trigger the system. As shown, the bandpass amplifier and the blanker share a common cathode resistor. When the blanker is in a nonconductive state, it has no effect on the bandpass amplifier, and the latter functions normally. When however, a gate pulse appears at the grid of the blanker, it causes conduction and increases the current flow through the common cathode resistor. Consequently, the bandpass cathode becomes more positive than its grid (high negative bias) and thus cuts off the bandpass tube.

The horizontal pulse obtained from the sweep output coincides with the occurrence of the blanking and burst timing; hence, only the video chroma signals are permitted to pass through the bandpass amplifier. A variable resistor at the output regulates the amount of signal applied to the chroma detectors, hence controlling the degree of color appearing on the screen.

9-6.   G — Y derivation

Early color receivers applied the generated 3.58-MHz subcarrier signals to I and Q demodulators (with the Q receiving a 90° phase lag subcarrier) where the missing carrier is combined with the sideband information and then detected to produce the color signals. Matrix systems were used to obtain $R - Y$, $G - Y$, and $B - Y$ signals. These, in turn, were matrixed with the luminous Y signal to obtain the final red, blue, and green signals.

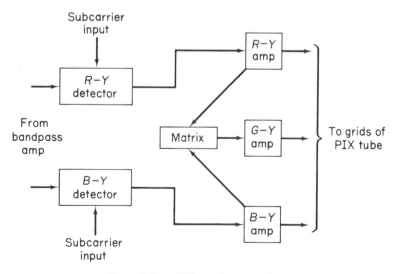

figure 9-9 *R-Y/B-Y* color detection

Modern receivers use the system shown in Fig. 9-9, wherein the demodulated signals are detected along the $B - Y$ axis and the $R - Y$ axis (see Fig. 9-2), with both signals having the same bandwidth and thus not requiring delay-line time-difference correction between I and Q. As shown, two demodulators again mix chroma signals with subcarrier signals, thus detecting $R - Y$ and $B - Y$, which are amplified by the respective stages, as shown. In addition, a resistive matrix network derives the $G - Y$ signal, which is also amplified. The resultant output signals are then applied to the grids of the color tube.

As shown in Fig. 9-6, the Y signal is applied to the picture-tube cathodes, and thus is electronically added to the color-difference signals applied to the grids. Essentially, therefore, the picture-tube elements function as a matrix for obtaining the required green, blue, and red signals. Usually both demodulators include low-pass filters for obtaining a 0.5-MHz bandpass to help reduce cross talk.

The phase relationships that apply for obtaining the color signals with $R - Y/B - Y$ (or I/Q) demodulation are shown in Fig. 9-10, which is part

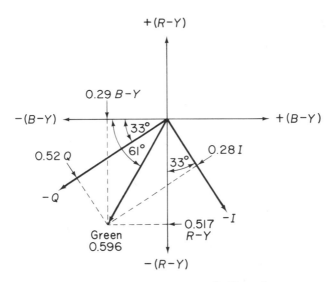

figure 9-10 either I/Q or R-Y/B-Y produce green

of the color wheel shown earlier in Fig. 9-2. Note that the $B - Y$ signals are displaced 33° clockwise from the Q signals, while the $R - Y$ signals are similarly displaced 33° clockwise, but from the I signals. Using green as an example, the signal voltages of either I/Q or $R - Y/B - Y$ will produce this color. Thus, if an amplitude of −0.52 is selected along the −Q axis and −0.28 along the −I axis, and these points are used to form a parallelogram, as shown by the dashed lines, the vector addition of the $-Q$ and $-I$ amplitudes results in 0.596, representative of the magnitude of the saturated (purity of color) green signal.

An identical amplitude for the green signal can be obtained by selecting an amplitude of 0.29 along the $-(B - Y)$ axis and -0.517 along the $-(R - Y)$ axis. A parallelogram formed from these new points produces identical vectors and the same amplitude obtained for the I/Q parallelogram.

9-7. killer and burst amplifier

To stop signals from entering the bandpass amplifier during black-and-white reception on a color set, the color-killer system shown in Fig. 9-11 is utilized. Note the linkage between the killer, the burst amplifier, bandpass amplifier, and phase detector. (See also Fig. 9-6.)

Note that the picture-signal input is applied to the burst-amplifier input and also coupled to the bandpass-amplifier grid. A positive potential is applied directly to the cathode of the burst amplifier, so conduction stops. (If the cathode voltage is made sufficiently high, the grid becomes relatively negative and reaches cutoff). During nonconduction, no signal voltage is applied to the phase detector from the burst amplifier.

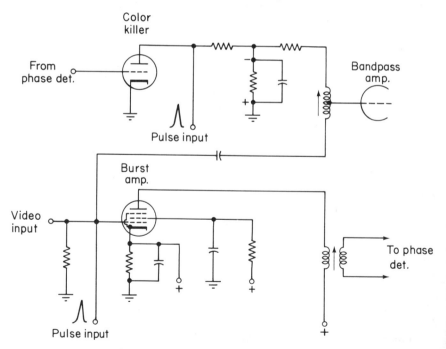

figure 9-11 color killer and burst amplifier

A positive pulse obtained from the horizontal output system is applied to the grid of the burst amplifier. Since such a pulse is synchronized with the horizontal sweep, it occurs only during the horizontal blanking interval. Thus, when the pulse appears at the grid of the burst amplifier, it causes tube conduction and, hence, permits the 3.58-MHz sync burst (riding on the blanking pedestal) to be applied to the phase detector of the subcarrier oscillator.

When the burst signal is applied to the phase detector, the latter applies a negative potential to the grid of the killer tube and cuts this tube off. A positive pulse from the horizontal-sweep section is also applied to the anode of the killer tube, as shown. Because this tube has no applied dc anode potential, the pulse substitutes for it. Since, however, the tube is held at nonconduction by the phase-detector potential, no current flow occurs throught the tube.

During black-and-white picture reception, there is no burst signal present on horizontal blanking; hence, the phase detector receives no such signal and thus no cutoff bias is applied to the killer-tube grid. The color killer tube now conducts, and the current flow through the shunting resistor sets up a voltage with a polarity that is negative toward the anode, as shown. This negative potential appears at the grid of the bandpass amplifier and has sufficient amplitude to cause the bandpass tube to cut off and no longer pass the video signal.

The capacitor across the shunting resistor filters the pulse voltage and converts it to a fairly steady amplitude dc voltage. The killer-tube function is controlled by a variable resistor in the input circuitry, and maximum signal coupling to the bandpass amplifier and phase detector is made by adjustment of the core position indicated by the arrow.

9-8.   3.58-MHz oscillator

The 3.58-MHz subcarrier crystal oscillator is shown in Fig. 9-12, plus the linkage with the reactance-control circuitry. The burst input is applied to the dual-diode phase detector and compared with the frequency and phase of the output signal from the oscillator. The output from the oscillator is also applied to the color demodulators, with precise tuning for maximum signal transfer made possible by the common-core variable-position slug between the transformer windings.

For the system shown in Fig. 9-12, control is accomplished by using a reactance tube circuit of the type discussed in Chapter 7 and illustrated in Fig. 7-6b. A slug-tuning adjustment is provided in the anode circuit of the reactance tube for precise control synchronization with the 3.58-MHz oscillator. Though the crystal stabilizes the subcarrier oscillator, a slight drift

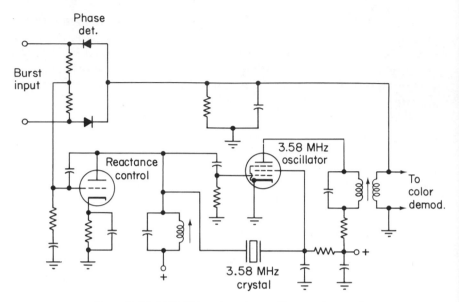

figure 9-12 3.58 MHz subcarrier oscillator and control

sufficient to cause phase differences could affect color rendition. Thus, the subcarrier oscillator is locked into exact synchronization with the original subcarrier at the transmitter by sampling the sync burst and using it for frequency control.

9-9. high voltage and sweep output

A typical output section of the horizontal-sweep system is shown in Fig. 9-13. For large-screen color picture tubes, an anode voltage of about 25,000 V is required for good screen brightness and beam acceleration. This potential is obtained by stepping up the amplitude of the pulse present in the horizontal-sweep output and rectifying it with the tube diode shown. As with the black-and-white receivers, the inner and outer conductive coatings of the picture tube form the filter capacitor for reducing ripple in the generated high voltage.

Since it is important to maintain good high voltage stability ın color receivers, a voltage-regulation system is employed. In some receivers this is done by using solid-state diodes, though the high-voltage shunt-regulator-tube system shown in Fig. 9-13 has been extensively used. A portion of the high voltage is sampled and applied to the grid of the shunt regulator tube. If the voltage declines below that set by the high-voltage adjusting potentio-

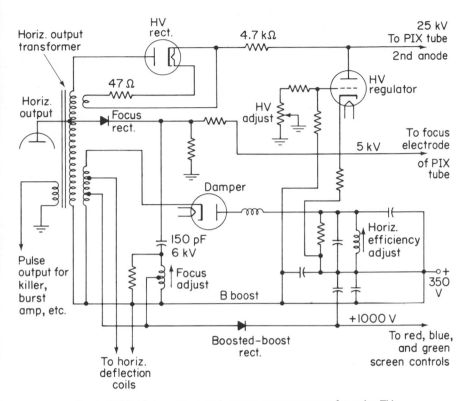

figure 9-13 high-voltage and sweep-output system for color TV

meter (in the regulator grid circuit), the voltage on the grid causes the regulator tube to conduct less. With reduced conduction there is less load on the high-voltage system and, in consequence, voltage rises to compensate for the decline. Similarly, if the high-voltage rises above normal, there is an increase in the positive potential applied to the grid of the regulator tube and greater conduction occurs. Thus, the increased shunt current drops the high voltage to the proper level.

A focus rectifier (often consisting of stacked solid-state diodes for higher voltage tolerance) supplies the necessary 5000 V of direct current for the focus electrode of the color picture tube. The voltage is adjustable by the slug-tuned inductor, as shown. The damper tube minimizes transient voltages developed in the system and boosts the low voltage. In addition, another rectifier is used to give the boosted voltage an additional boost, bringing it to 1000 V or more for color screen control voltages when required. A tap on the secondary winding of the horizontal output transformer (plus the bottom of the winding) supplies the sweep signals for the horizontal-deflection coils, as shown.

questions and problems

9-1. How was it possible to add color-information signals to the already crowded television-station spectrum? Explain briefly.

9-2. Explain briefly why the subcarrier *frequency* of 3.579545 MHz was chosen.

9-3. What are the exact frequencies of the vertical and horizontal sweep rates in color transmission?

9-4. What proportions of red, blue, and green signals make up the Y signal? Briefly explain the reason for using such relative values.

9-5. What relative amplitudes of color signals make up the I and Q signals in transmission?

9-6. In what manner is the color subcarrier eliminated at the transmitter after the color sidebands have been produced? Explain briefly.

9-7. What bandwidths are assigned to the I and Q signals?

9-8. Explain briefly what causes a difference in time coincidence in the Y, I, and Q signals and what is done to correct it.

9-9. What is the frequency of the burst signal present on the horizontal pedestal during color transmission, and what is its purpose?

9-10. Why is black-and-white reception impaired if one gun of a three-gun color picture tube becomes defective?

9-11. Why is a separate sound-IF detector used in color receivers? Explain briefly.

9-12. Why is the Y signal delayed with respect to the I and Q in a television color receiver?

9-13. Briefly explain the purposes for the bandpass amplifier and the blanker circuits in a color-television receiver.

9-14. In what manner is the $G - Y$ signal obtained in a color receiver?

9-15. What are the functions of the color killer and burst amplifier? Explain briefly.

9-16. Using a block diagram, show how the subcarrier oscillator in a color receiver is kept at the exact frequency of the original subcarrier at the transmitter.

9-17. What method is used to reduce ripple in the 25,000 V applied to the second anode of a color tube?

9-18. How is voltage stability maintained in the high-voltage system of a color receiver?

9-19. What method is used for obtaining an additional increase in the boosted low voltage?

9-20. How is the necessary high potential obtained for focus purposes in a color receiver? Illustrate your explanation by a representative drawing.

10

filters and transmission lines

10-1. introduction

Various filter networks are used in communications for removing or passing only signals of certain frequencies as required to achieve a specific type of transmission. In some instances only low-frequency signals are to be passed, while on other occasions it may be necessary to pass signals having only high frequencies. When only a narrow band of signals is required, filters of the bandpass variety are used. If, on the other hand, a band of signals is to be removed but other signals above and below the band in frequency are to be retained, band-stop filters are convenient. Thus, reactive or resonant circuitry finds wide applications in communication systems for modifying various signal groups.

Transmission lines have many uses, among them the transfer of signals between various units (such as transmitter and antenna, or antenna and receiver), as reactive or resonant sections at high-frequency use, and as signal-delay devices. Thus, filter sections (as well as resonant circuits for amplifiers and oscillators) often take the form of short sections of line at the very high or ultrahigh frequencies, where ordinary capacitors and inductors are no longer useful.

If the resonant frequency is to be increased in a circuit composed of inductance and capacitance, either the value of the capacitor or the inductor must be decreased. Eventually, however, the capacitor becomes extremely small and may only have two plates. Similarly, the inductor may become only a single loop with a small diameter. Since, however, sections of transmission lines exhibit capacitive and inductive characteristics, they become very useful at frequencies that are too high for the ordinary physical components of inductance and capacitance.

Initially, various filters are discussed in this chapter for an understanding of their characteristics and design. Next, transmission lines and sections thereof are covered, with an analysis of the factors and design procedures producing equivalent units of inductance and capacitance for forming filters, resonant tuning circuitry, etc.

This chapter also serves as an introduction to the microwave principles covered in Chapter 11. For the microwave region, transmission units take the shape of metallic and hollow rectangular or circular pipes. Again, as with the lines discussed in this chapter, short sections are used to obtain characteristics of inductance, capacitance, or combinations of the two as needed.

10-2. low-pass (k) filters

Low-pass filters permit the transfer of lower-frequency signals through a system while attenuating or eliminating signals of higher frequencies. The basic low-pass filter circuit is shown in Fig. 10-1a, and contains a series inductor L_1 followed by a shunt capacitor C_1. If various signals of different

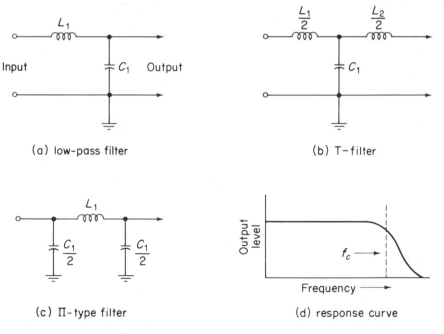

(a) low-pass filter   (b) T-filter

(c) Π-type filter   (d) response curve

figure 10-1 low-pass constant-k filters

frequencies are applied to the input of this filter, signals having progressively lower frequencies encounter decreased inductive reactance for L_1 and will pass through the filter. Capacitor C_1 has a high reactance for lower-frequency signals and, hence, offers little shunting effect.

Signals having progressively higher frequencies meet with higher inductance reactance and will be diminished at the output of the filter. Also, for higher-frequency signals the capacitor has a much lower reactance and, hence, also shunts some of the signal energy.

The low-pass filter of Fig. 10-1a resembles an inverted capital letter L; hence, it is sometimes referred to as an L-type filter. It is also known as a *half-section*, and is usually used with added components for more effective filtering purposes, as shown in Fig. 10-1b, where an added inductor is present. This type is also known as a T filter, since it resembles the capital letter T. The second inductor presents an additional series reactance for low-frequency signals.

For the filter of Fig. 10-1c, an additional shunt capacitance is used at the input, providing additional attenuation for high-frequency signals. The upper inductance, with a shunt capacitor each side, makes the network appear as the Greek letter pi (π); hence, this filter is often termed a π-type filter. The bandpass characteristics of such a filter depend on the reactive values selected for the frequencies of the signals to be attenuated, plus the number of half-sections employed, as more fully discussed later. Generally, a graphed response curve is as shown in Fig. 10-1d, where the cutoff frequency (f_C) indicates the frequency of the signal above which attenuation occurs. Often the f_C refers to the 70 per cent of peak value.

10-3. filter factors

All filters composed of reactive components have a constant impedance that remains unchanged even though additional half-sections are added or the signal voltage amplitude changes. Thus, if the low-pass filter section of Fig. 10-1a had an infinite number of similar sections connected to it in progressive order, the orignal impedance of the circuit would remain the same. This is so because no resistive components were present originally or added when sections were connected to it (assuming pure inductances and capacitances). Since neither an inductor nor capacitor consumes electric energy, no change occurs in the impedance of the filter sections.

With an infinite number of sections, the filter assumes the electrical configuration of a transmission line, as shown in greater detail later. Electrical energy applied would cause current to travel through the successive half-sections, as the components of L and C charged and discharged. If an ammeter were placed in series with the line made up of an infinite number of sections, a specific current value would be read as energy travels through the infinitely long line. Since $E/I = Z$ (or R), we obtain an ohmic value indicating the inherent circuit impedance (Z_o), which is termed the *characteristic impedance* of the system, and also refers to a transmission line used for transfer of electric energy between two points. (Also note that for Fig. 10-1a, the addition of each series inductive reactance is nullified by the inclusion of a capacitive-reactance shunt as the system is extended. Again, Z_o remains constant.)

When any filter composed of one or more half-sections of L and C is terminated in a resistance equal to the characteristic impedance (Z_o), signal current flow equals that for the infinite line, and again $E/I = Z_o$. The characteristic impedance is also known as *iterative* (repetitive) *impedance*, and also *surge impedance*. A maximum signal energy transfer through the system is only obtained when the load resistance applied to the output matches the characteristic impedance. The characteristic impedance is determined by

$$Z_o = \sqrt{\frac{L}{C}} \qquad (10\text{-}1)$$

For the low-pass filters shown in Fig. 10-1 the term *constant k* is also used; it refers to a symmetrical system in which the product of the series and shunt reactances remains fixed in value regardless of the frequency of applied signal. Thus, for the low-pass filter of Fig. 10-1a we can designate the series reactance as Z_1 (to include any possible resistive components) and the shunt reactance as Z_2. Hence, we obtain the following characteristic:

$$Z_1 Z_2 = k^2 \qquad (10\text{-}2)$$

The k factor is constant for all signal frequencies, and Eq. (10-2) also applies to Fig. 10-1b or c. Total inductance values can be ascertained by

$$L = \frac{R}{\pi f_c} \qquad (10\text{-}3)$$

where R = terminating resistance
f_c = cutoff frequency in hertz
L = the inductance in henrys

Note that in Fig. 10-1b each inductor is identified by $L_1/2$ because each (in series) has half the value of the total inductance. Similarly, in Fig. 10-1c each capacitor is indicated as $C_1/2$, because each capacitor in shunt contributes to one half the total capacitance value. Total capacitance is found by

$$C = \frac{1}{\pi f_c R} \qquad (10\text{-}4)$$

where R = terminating resistance
f_c = cutoff frequency in hertz
C = total capacitance in farads

The following equation applies for finding the cutoff frequency for the constant-*k* low-pass filter:

$$f_c = \frac{1}{\pi\sqrt{LC}} \qquad\qquad (10\text{-}5)$$

10-4.   low-pass m-derived filters

An *m-derived* filter is a special type that produces a sharper and more defined cutoff by the addition of another component (in series or shunt) to the basic constant-*k* type. A typical example is shown in Fig. 10-2a where

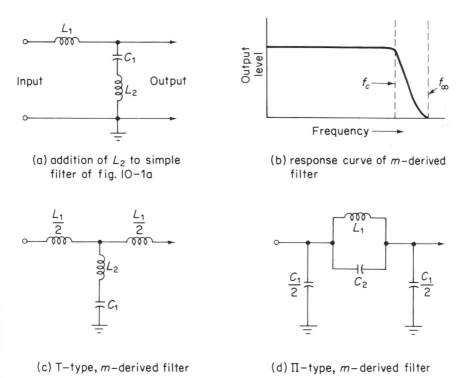

(a) addition of L_2 to simple filter of fig. IO-1a

(b) response curve of *m*-derived filter

(c) T-type, *m*-derived filter

(d) Π-type, *m*-derived filter

figure 10-2 low-pass *m*-derived filters

an inductance L_2 has been added to the original half-section low-pass filter of Fig. 10-1a. Because the addition of L_2 places it in series with C_1, a series-resonant circuit is formed for a particular frequency. At the resonant frequency the impedance of the series-resonant circuit is low, and this characteristic can be used to produce a sharp cutoff in the filter curve.

The *m*-derived filter thus utilizes the resonance of the circuit to obtain infinite attenuation at a specific frequency beyond the cutoff frequency f_C. The component impedances are interrelated by a constant *m*, which, in equation form, is related to the ratio of the cutoff (f_C) and the frequency of infinite attenuation (designated as f_∞).

The value of *m* is fractional and generally is approximately 0.6. For a sharper cutoff the *m* value is set nearer 0. The equation for the *m* value for a low-pass filter is

$$m = \sqrt{1 - \left(\frac{f_C}{f_\infty}\right)^2} \qquad (10\text{-}6)$$

The response curve of Fig. 10-2b shows the degree of attenuation obtained with the *m*-derived filter. For *m* values below 0.5, a sharper decline is obtained between the f_C and f_∞ points.

The *m*-derived filter shown in Fig. 10-2c is also known as a T type since it resembles that capital letter. Again the shunt components form a series-resonant circuit. In Fig. 10-2d the two shunting capacitors form a π-type filter (*m*-derived) and the additional component is a capacitor, C_2, shunting L_1, hence forming a parallel-resonant circuit. Since the parallel-resonant circuit offers a high impedance for the signals of resonant frequency, a sharp attenuation is again provided by utilizing the resultant dip formed in the response curve.

For the *m*-derived low-pass filter of Fig. 10-2a, the L_1 and C_1 values are found by

$$L_1 = \frac{mR}{\pi f_C} \qquad (10\text{-}7)$$

$$C_1 = \frac{m}{\pi f_C R} \qquad (10\text{-}8)$$

10-5. high-pass k filter

As the name implies, the purpose for the high-pass filter is to pass signals having frequencies above a specific one, while diminishing signals of lower frequencies. Since the function is opposite to that of the low-pass filter, the *L* and *C* components are interchanged as shown in Fig. 10-3a, where the capacitor is now in series (instead of shunt as shown in Fig. 10-1a) and the inductor is across the output. Thus, signals with progressively higher frequencies meet with decreased capacitive reactance for C_1 and appear at the output. For lower-frequency signals, however, there is an increasing capacitive reactance in series and attenuation results. Similarly, lower-frequency signals are shunted by the low reactance of inductor L_1, but the higher-fre-

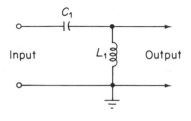

(a) simple high-pass filter

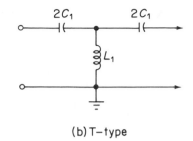

(b) T-type

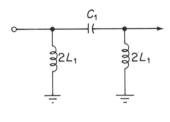

(c) Π-type

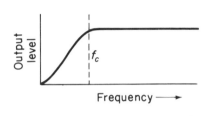

(d) response curve

figure 10-3 high-pass constant-*k* filters

quency signals find an increasingly higher inductive reactance and, hence, suffer less attenuation.

As with the low-pass filters, for a maximum transfer of signal energy between the input and output, the impedance Z_o of the filter section must match that of the terminating (load) resistance. The factors relating to constant *k* discussed for the low-pass filter also apply to the high-pass type.

The T-type high-pass filter is shown in Fig. 10-3b, and the π type in Fig. 10-3c. The high-pass filter characteristics are graphed in Fig. 10-3d with the cutoff frequency (f_c) now indicating the frequency *below* which attenuation occurs. As with the low-pass filter, the characteristic impedance if found by Eq. (10-1).

Appropriate equations (with unit values the same as for the low-pass filter types discussed earlier) are

$$L = \frac{R}{4\pi f_c} \tag{10-9}$$

$$C = \frac{1}{4\pi f_c R} \tag{10-10}$$

$$f_c = \frac{1}{4\pi\sqrt{LC}} \tag{10-11}$$

10-6.   high-pass m-derived filter

As with the low-pass filters, the *m*-derived sections can also be used with the high-pass constant-*k* types for obtaining a sharper decline around the cutoff point. The various basic types are shown in Fig. 10-4, with the series capacitor C_2 added to form a low impedance for the resonant frequency selected. Now the cutoff point (f_c) and the point of infinite attenuation (f_∞) show the response decline, as in Fig. 10-4b.

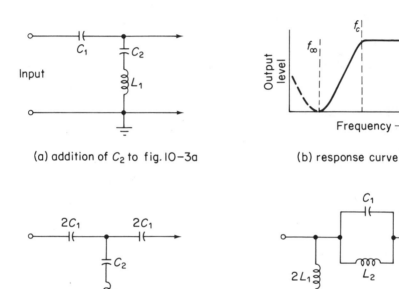

(a) addition of C_2 to fig. 10–3a   (b) response curve

(c) T–type, *m*–derived filter   (d) Π–type, *m*–derived filter

figure 10-4 high-pass *m*-derived filters

For the π-type shown in Fig. 10-4d, the addition of L_2 forms the parallel-resonant circuit, which presents a high impedance for signals having frequencies around the resonant frequency. For the high-pass filters, the *m* value is found by

$$m = \sqrt{1 - \left(\frac{f_\infty}{f_c}\right)^2} \qquad (10\text{-}12)$$

For Fig. 10-4a, the component values (L_1 and C_1) are found by

$$L_1 = \frac{R}{4\pi f_c m} \qquad (10\text{-}13)$$

$$C_1 = \frac{1}{4\pi f_c m R} \qquad (10\text{-}14)$$

The added component C_2 of Fig. 10-4a and c is found by

$$C_2 = \frac{m}{(1 - m^2)\pi f_c R} \qquad (10\text{-}15)$$

The shunting inductor L_2 for the parallel-resonant circuit of Fig. 10-4d is found by

$$L_2 = \frac{mR}{(1 - m^2)\pi f_c} \qquad (10\text{-}16)$$

10-7.   *balanced filters*

All the filters discussed have been the *unbalanced* type in which one line is operated at ground potential. With one line grounded, the other line is above ground or *unbalanced* with respect to ground. All the filters previously discussed can be balanced with respect to ground, as shown by the two examples in Fig. 10-5.

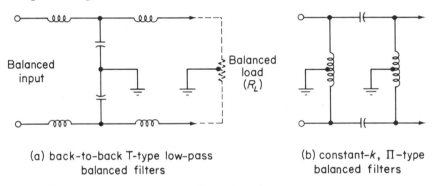

(a) back-to-back T-type low-pass balanced filters	(b) constant-k, Π-type balanced filters

figure 10-5 balanced filters

In Fig. 10-5a, two T-type filters (low-pass) are connected back to back. This arrangement permits placing the center portion of the composite filter at ground, as shown, with the upper and lower lines now balanced with respect to ground. Similarly, the constant-k π-type high-pass filter can be balanced with tapped inductors, as shown in Fig. 10-5b.

10-8.   *bandpass filters*

A bandpass filter has the ability to transfer from its input to output a desired group of signals that have frequencies around the resonant one. The width of the frequency span is determined by the selectivity (Q) of the component combination. Undesired signals having frequencies above and below the selected passband will be attenuated or filtered to negligible amplitudes.

Representative bandpass filters are illustrated in Fig. 10-6a, c, and d—,

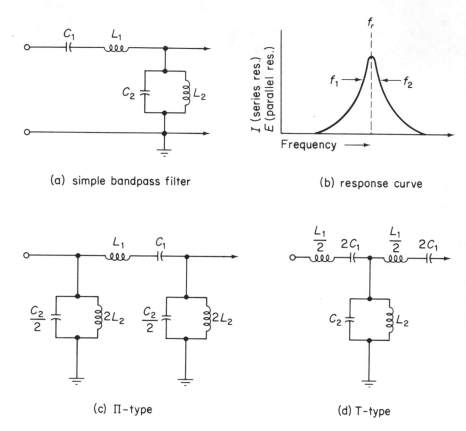

(a) simple bandpass filter

(b) response curve

(c) Π-type

(d) T-type

figure 10-6 bandpass filters

an L type, a π type, and a T type, respectively. For the L type both the series-resonant circuit (C_1 and L_1) as well as the parallel-resonant circuit (C_2, L_2) are tuned to the resonant frequency around which the bandpass is to occur. Thus, signals having frequencies at and near the resonant one encounter a low impedance in the series-resonant circuit and are passed through the filter. For the signals desired, the parallel-resonant circuit shunting the output has a high impedance and, hence, offers negligible attenuation.

For signals having frequencies above or below the selected resonant frequency, the series-resonant circuit offers a high impedance; hence, attenuation is high. Signals arriving at the output are shunted by the parallel-resonant circuit, since this network has a low impedance for signals of frequencies off resonance.

A typical selectivity curve for the bandpass filter is shown in Fig. 10-6b. Resonant-frequency factors and Q are the same as for resonant circuits discussed in Chapter 3. Thus, the resonant frequency (f_r) for either a series- or parallel-resonant circuit is determined by using Eq. (3-4). The circuit Q in relation to the frequency points shown is found by using the appropriate equations for the series or the parallel circuit, as covered in Section 4-3.

Individual values for the bandpass filters shown in Fig. 10-6 are determined by using the following equations:

$$L_1 = \frac{R}{\pi(f_2 - f_1)} \tag{10-17}$$

$$L_2 = \frac{(f_2 - f_1)R}{4\pi f_1 f_2} \tag{10-18}$$

$$C_1 = \frac{f_2 - f_1}{4\pi f_1 f_2 R} \tag{10-19}$$

$$C_2 = \frac{1}{\pi(f_2 - f_1)R} \tag{10-20}$$

10-9. bandstop filters

A bandstop filter is sometimes known as a band elimination filter, since it eliminates a group of signals having frequencies around the resonant one, but passes signals of frequencies above and below the resonance frequency. The width of the band to be eliminated is set by the Q of the resonant circuits used, as with the bandpass filters.

Typical bandstop filters are shown in Fig. 10-7a, c, and d—, the L type, the π type, and the T type, respectively. For the constant-k type of Fig. 10-7a, the parallel-resonant circuit composed of C_1 and L_1 offers a high impedance at resonance, hence attenuating signals at or near the resonant frequency. Also, the series-resonant circuit shunting the output (C_2, L_2) has a low impedance for the signals to be eliminated, hence offering additional attenuation.

For signals having frequencies above and below the resonant one, the parallel-resonant circuit provides a low impedance, thus passing such signals. The desired signals, being off resonance, find a high impedance in the series circuit shunting the output and suffer negligible attenuation.

The bandstop curve is shown in Fig. 10-7b, with f_1 representing the low-frequency end of the bandstop span, and f_2 the high-frequency end. This is an inverted selectivity curve, and the width of the bandstop is again determined by the Q factors for resonant circuits covered in Chapter 4, as with the bandpass filter.

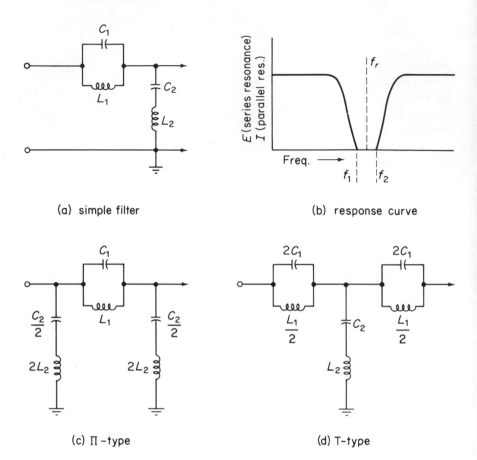

(a) simple filter

(b) response curve

(c) Π-type

(d) T-type

figure 10-7 bandstop filters

The values for the various components of the bandstop filter are found by

$$L_1 = \frac{(f_2 - f_1)R}{\pi f_1 f_2} \tag{10-21}$$

$$L_2 = \frac{R}{4\pi(f_1 - f_2)} \tag{10-22}$$

$$C_1 = \frac{1}{4\pi(f_2 - f_1)R} \tag{10-23}$$

$$C_2 = \frac{f_2 - f_1}{\pi R f_1 f_2} \tag{10-24}$$

10-10. transmission lines

Four basic transmission-line types used in communication systems are shown in Fig. 10-8. Figure 10-8a is the *open-wire* line wherein two con-

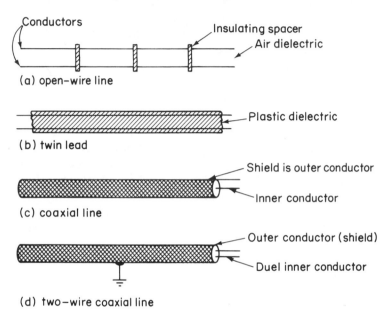

figure 10-8 transmission line types

ductors are spaced by plastic or ceramic insulators. When any metal objects or wires are brought into close proximity, capacitance exists between them. Thus, the term *dielectric* is used, just as with standard capacitors, for the air or insulation material within the spacing. Thus, for the open-wire line, the dielectric is air. Such a line is characterized by low loss, since dielectric materials such as plastic usually introduce some shunt resistance that consumes signal energy. The disadvantage is the wider spacing for achieving the same impedances as with plastic-dielectric lines, plus greater radiation losses.

The plastic-dielectric two-wire line shown in Fig. 10-8b is the common *twin lead* used in television receiver installations. Such a line is standardized at 300-Ω characteristic impedance, and the flexible insulation and less than half-inch spacing between wires provides for convenient installation and feed through walls and other spaces. Losses are somewhat greater than for the open-wire line, and some signal (interference) pickup may occur along such a line.

The transmission line of Fig. 10-8c is referred to as a *coaxial cable* or *concentric line*. The inner conductor (of solid or stranded wire) is held at center by a flexible plastic insulation or by washer-like dielectric insulators at fixed points within the transmission line. The outer conductor may be a metal tube or metallic braid to provide for flexibility. This outer conductor is usually placed at ground potential, thus forming an unbalanced line, just as in the case of the unbalanced filter sections discussed earlier. The coaxial cable, because of the outer conductor shielding the inner, is superior to other line types in terms of minimum radiation losses or interference pickup. Signal attenuation, however, is generally higher.

The coaxial cable finds some applications in receiver installations where connections must be made to a remote antenna with a minimum of noise pickup. This cable finds extensive use in transmitting systems for interconnection of microphones, studio equipment, transmitter output to antenna, ultra-high frequency filter networks, and similar applications. For balanced systems, the two-wire coaxial line shown in Fig. 10-8d is used, with the shielded outer conductor at central ground potential with respect to the dual inner lines.

Figure 10-9a shows the electrostatic lines of force existing between the

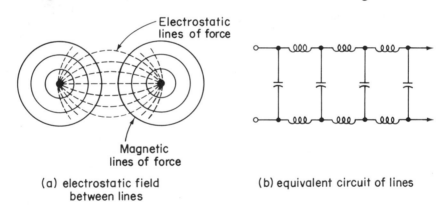

(a) electrostatic field
between lines

(b) equivalent circuit of lines

figure 10-9 line characteristics

two current-carrying wires of a transmission line, plus the magnetic lines of force surrounding them. The amplitude of such fields depends on the amount of current flow and the applied potential. Such fields exist for either direct or alternating current, though for the latter the lines of force collapse as each ac alternation drops to zero and builds up again (in opposite polarity) as the next ac alternation reaches a peak amplitude.

As shown in Fig. 10-9b, each transmission line has a specific amount of inductance per unit length of line (as does any length of wire). Since close proximity of two wires forms capacitance, a certain amount is also present per unit length, except it is represented as in shunt instead of in series as with inductance. Some resistance is also present in the wire, though this is kept at a negligible value by using larger-size conductors. For dielectric materials other than air, some shunt resistance may also occur, particularly if the leakage resistance of the dielectric material is impaired by moisture or inherent impurities.

As with the filter sections discussed earlier, the repeated units of inductance and capacitance along the line set up a characteristic impedance; hence, Eq. (10-1) also applies. For a given transmission line, the impedance remains the same regardless of the line length. Thus, if a 300-Ω twin lead is 10 ft or 100 ft long, the Z_o remains at 300 Ω.

When air insulation is used in a parallel-wire line, such as in Fig. 10-8a, the following equation can be used for Z_o:

$$Z_o = 276 \log \frac{2b}{a} \qquad (10\text{-}25)$$

where b = spacing (center-to-center) between wires
a = diameter of each conductor

Since b/a is a ratio, values can be in inches, feet, millimeters, etc., and the Z_o value will be the same. For values in inches, the following is a typical example:

A two-wire (air-dielectric) transmission line was measured and the diameter of each wire was found to be 0.04 in. Spacing between the wires was 3 in. What is the characteristic impedance?

$$\textit{Solution:} \quad Z_o = 276 \log \frac{2 \times 3}{0.04}$$
$$= 276 \log 150 = 276 \times 2.1761$$
$$= 600 \ \Omega$$

If the values of L and C were known, the same impedance would be found by using Eq. (10-1):

An air-dielectric transmission line had an inductance value of 0.36 mH, and a capacitance of 0.001 μF for a measurement of 75 ft. What is the impedance?

$$\textit{Solution:} \quad Z_o = \sqrt{\frac{L}{C}} = \sqrt{\frac{0.36 \times 10^{-3}}{10^{-9}}}$$
$$= \sqrt{0.36 \times 10^6}$$
$$= 0.6 \times 10^3$$
$$= 600 \ \Omega$$

If the inductance and capacitance values were taken for a longer section of line, the surge impedance value would remain the same, since for each increase in series inductance, there would be a corresponding increase in shunt capacitance, thus maintaining the same ratio of L/C.

For a coaxial cable, such as in Fig. 10-8c, the characteristic impedance may be found by

$$Z_o = 138 \log \frac{b}{a} \tag{10-26}$$

where $b =$ inside diameter of the outer conductor
$a =$ outside diameter of the inner conductor

When the coaxial cable's dielectric material is other than air, Eq. (10-26) should be multiplied by the results of the formula

$$\frac{1}{\sqrt{k}} \tag{10-27}$$

where k is the *dielectric constant* of the material.

When discussing transmission-line characteristics, the general practice is to term the signal source the *generator* and the signal recipient unit the *load*, where power is consumed or radiated (as in transmitting antennas). Such designations are convenient for dispensing with the repeated reference to the transmitter, the receiver, the antenna, etc. Thus, a generator could be a transmitter with an antenna as the load, while in the receiver the antenna is the generator and the receiver the load.

As with the filter sections discussed earlier and other generator–load combinations, the maximum available signal power is transferred between generator and load only when the impedance of one matches the other.

As shown later, transmission lines can be utilized in such a manner that advantage is taken of the inherent resonant characteristics. If, however, the resonant factors are ignored, the line impedance must match that of the generator and load for a maximum signal-energy transfer. In such an instance, the transmission line is termed a *flat* line, or *untuned* line. Such a non-resonant line is "flat" regarding power values along the line, since voltage and current values remain unchanged, as does impedance. Such is not always the case, as discussed more fully later for the resonant-type line.

For the ac signal applied to a transmission line, the lines of force, shown in Fig. 10-9a, are equal but opposite for the two wires, since current flow at any instant is in one direction for a particular wire and in the opposite direction for the other wire. Thus, the fields tend to oppose each other and cancel, depending on the proximity of the wires. Hence, radiation losses are reduced to a greater degree as the wires are brought closer together. Since, however, spacing alters Z_o [Eq. (10-25)], the wires may have to be farther apart than desirable for minimum losses.

10-11. line-length factors

The study of transmission-line factors necessitates repeated references to the specific wavelength of the line, since line length relates to inductive, capacitive, or resonant characteristics. (Section 1-2 should be reviewed, and reference made to Fig. 1-1.)

When a transmission line's Z_o matches the generator's Z, the maximum amount of available signal energy enters the line and travels toward the load terminating the line. If, however, the load is not matched to the line, all the signal energy is not accepted and some reflects back to the generator. The greater the mismatch, the more energy returned along the line. Hence, some signal energy travels along the line toward the load, while some is also returning along the line back to the generator.

Signal travel in both directions along the transmission line causes the primary signals and the reflected signals to be intermixed along the line, resulting in out-of-phase conditions in some sections and in-phase conditions at other sectors along the line. Consequently, at places where voltages are in phase their amplitudes are high, while at sections where out-of-phase conditions prevail, low or zero voltages occur. Similar high and low amplitudes prevail for current values along the line where reflections are present.

The high and low amplitude points of voltage and current along the line remain at fixed positions, and, hence, are referred to as *standing waves*. High points of voltage or current are termed *loops*, while low or zero amplitude points are called *nodes*. Since voltage and current amplitudes vary along a line with standing waves, the line impedance no longer is constant, but depends on the voltage–current ratio prevailing at any particular point.

An extreme condition of mismatch prevails if the transmission line is either open or shorted at the terminating end away from the generator. In such instances all the signal energy is returned to the generator, since no load resistance is present to consume the energy. With an open line the impedance is infinitely high, while the shorted (closed) line drops the impedance to zero. The voltage and current relationships form out-of-phase conditions, as shown for the various lengths of lines (both open and shorted) of Fig. 10-10.

Figure 10-10a shows a line that is a half-wavelength long and open ended. Current and voltage distribution along such a half-wavelength line result in a voltage loop at the end, plus a current node. When the wave of signal energy reaches the end of such a line, the current drops to zero, and the resultant collapsing fields cut the conductor ends and induce a voltage having a peak value, as shown, with a high impedance. For a half-wavelength line, a high-Z condition also prevails at the generator (though the voltage loop here is 180° out of phase with that at the end of the line, as is the current node out of phase with that at the line termination). While the loops and

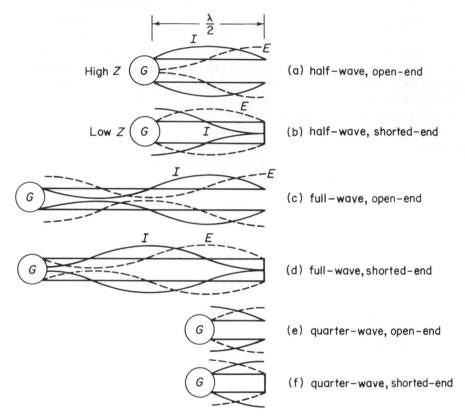

figure 10-10 open- and closed-line factors

nodes "stand" at fixed points, the energy is of ac composition; hence, all amplitudes of voltage and current repeatedly build up and collapse.

Figure 10-10b shows a line of the same length, but closed at the far end from the generator. The shorted condition creates a current loop at the end with zero voltage. Thus, the impedance is zero and again no matching load resistance is present, thus producing the standing-wave condition shown. If a load resistance were used instead of a short, the ratio of loop amplitude to node amplitude would change to the degree depending on whether the value of the load resistance approaches that of the generator impedance. The ratio of voltage loops and nodes (or current loops and nodes) is known as the *standing-wave ratio* and is found by dividing the maximum value of voltage along the line by the minimum voltage value. Such a standing-wave ratio thus indicates the degree of mismatch between the load resistance and the impedance of the generator.

When a complete match exists between load and generator, the standing-wave ratio is equal to 1, since loops and nodes no longer exist for either voltages or currents. Line reflections, however, cause standing waves and indicate mismatching. As an example, if the standing wave of current attains a value of 0.4 A and the minimum standing wave of current is 0.04 A, the ratio is 10. This indicates that the generator impedance is either 10 times larger than the load resistance or one half the value of the load resistance. Under these particular conditions, the standing-wave ratio of the voltage would, of course, also be 10.

Transmission-line sections a full wavelength long are shown in Fig. 10-10c and d. The open and closed terminations establish the same voltage and current relationships as the lines in Fig. 10-10a and b. The impedance at the generator of Fig. 10-10c, is high the same as at the end, while for Fig. 10-10d a low impedance prevails both at the generator and the termination, the same conditions that prevail in Fig. 10-10a and b. Thus, half-wavelength sections, or multiples thereof, can be used as one-to-one transformers because of identical input and output impedances. (Practical applications will be shown later.)

For the line sections of Fig. 10-10e and f, quarter-wavelengths are shown. For the open line of Fig. 10-10e, a voltage loop is present; hence, the impedance here is higher than at the generator. Consequently, such a length of line can be used as an impedance step-up device to match a low-impedance generator to a higher impedance load resistance unit. In Fig. 10-10f the closed termination drops voltage to zero but raises current. At the generator, however, voltage is high; hence, impedance is high. Thus, such a line section is useful as an, impedance step-down transformer.

10-12. resonant and reactive sections

Because the transmission-line sections have characteristics of series inductance and shunt capacitance, resonance must prevail for a specific frequency, as with any inductor–capacitor combination. Consequently, when the inductive- and capacitive-reactance values are equal (and opposite), resonance is achieved. It will be found, however, that the line is exactly *one quarter-wavelength* long for the frequency at which resonance is obtained.

For the quarter-wavelength line of Fig. 10-10e, the generator sees a series-resonant circuit because the impedance of the line at the point where it connects to the generator has a minimum of impedance, just as with a series-resonant circuit. For the quarter-wavelength resonant section of Fig. 10-10f, however, the generator connects to a high-impedance section; hence, the transmission line section has all the characteristics of a parallel-resonant circuit. As shown later, such sections replace conventional capacitor–inductor combinations for higher-frequency operations.

For the quarter-wave open-line resonant section the series resonance is destroyed if the line is either shortened or lengthened slightly. If, for instance, the line is shortened to less than a quarter-wavelength, both the inductance and capacitance values would decrease. Now, inductive reactance drops, but capacitive reactance rises. Since the quarter-wavelength open line is equivalent to a series circuit, the rise in capacitive reactance cancels the decreased inductive reactance, causing a predominance of capacitive reactance and, hence, capacitance. Thus, for the shortened quarter-wave section, the line behaves as a pure capacitance (assuming negligible resistance in the wires).

If the quarter-wave section of open line were made longer than a quarter-wavelength (but not to a half-wavelength), both inductance and capacity increase. Now the series opposition becomes primarily inductive reactance, forming a section equivalent to inductance. Since the shorted quarter-wavelength lines have impedance characteristics opposite to the open lines, an inductance is formed for a line decreased in length, and a capacitance is obtained when the line length is increased over the quarter-wavelength (but kept at less than a half-wavelength). When the line is now shortened, the lower inductance reactance in the parallel-resonant circuit carries most of the signal current, causing the current to have inductive characteristics. As open or closed sections of lines are increased in length beyond the quarter-wavelength, reactive characteristics invert, as shown in Fig. 10-11.

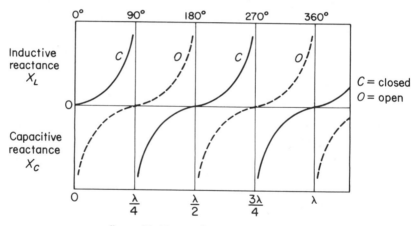

figure 10-11 reactive charges along line

10-13. practical applications

The term *stub* is applied to a short length of transmission line used for obtaining characteristics of inductance, capacitance, or resonance at high-frequency operation. The use of stubs for forming filters is shown in Fig. 10-12. In Fig. 10-12a a typical constant-*k* low-pass filter section is shown,

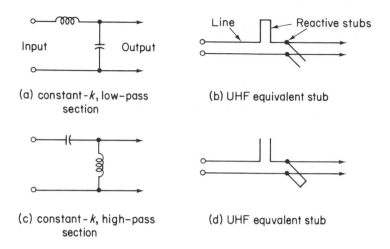

(a) constant-k, low-pass
section

(b) UHF equivalent stub

(c) constant-k, high-pass
section

(d) UHF equvalent stub

figure 10-12 stubs form filters

while Fig. 10-12b is the ultrahigh-frequency counterpart. Here, a closed-line section is placed in series with one leg of the transmission line. Such a stub is cut for a length less than a quarter-wavelength and, hence, behaves as an inductance and performs the same reactive function as its physical counterpart in Fig. 10-12a.

To form the shunt capacitor at the output of the filter, an open section of line (less than a quarter-wavelength long) is placed across the two wires of the transmission line (Fig. 10-12b), thus providing the necessary capacitance. Similarly, open and shorted stub sections also form the low-pass filter, as shown in Fig. 10-12c (low-frequency version) and Fig. 10-12d (high-frequency counterpart). The open stub, less than a quarter-wavelength, is again used as a capacitance, only now it is placed in series with one line only. The closed stub, acting as an inductance, is placed across the line to simulate the shunt inductance at the output of the high-pass filter.

If the stubs of Fig. 10-12b and d were made exactly a quarter-wavelength long, resonant filters would be obtained. In Fig. 10-12b the closed quarter-wave section in series with one line acts as a parallel-resonant circuit, hence offering a high impedance to a band of frequencies clustered around the resonant frequency. The open quarter-wave section across the line, however, acts as a series-resonant circuit, shunting signals at or near the resonant frequency. Thus, resonant stub sections form a bandstop filter with characteristics similar to the lower-frequency-spectrum version given earlier (see Fig. 10-7). Quarter-wave sections for the stubs of Fig. 10-12d would form a bandpass filter similar to the types shown in Fig. 10-6. The open stub would act as a series-resonant circuit, passing signals with frequencies at or near the resonant one.

The vestigial-sideband filter is a typical example of the use of line stubs for forming equivalent capacitance, inductance, and resonance. The required circuitry (Fig. 10-13) partially suppresses the lower sidebands of the television

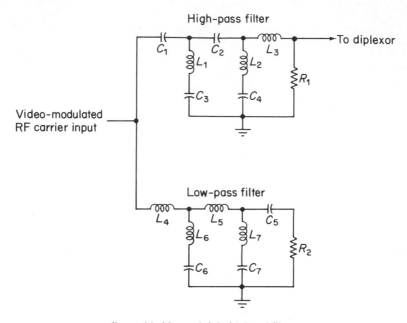

figure 10-13 vestigial-sideband filter

signal, as described in Section 8-2. As shown, high-pass and low-pass sections are used, with the undesired lower sidebands passing through inductors L_4 and L_5 and being shunted by the series-resonant cricuits composed of C_6, L_6, and L_7, C_7. Resistor R_2 is coupled to the low-pass filter by C_5 and terminates the filter section to eliminate line reflections.

Desired signals (the higher-frequency sidebands) find a low reactance in C_1 and C_2 and, hence, enter the high-pass filter. Inductor L_1 plus capacitor C_3 as well as L_2 and C_4 have a low shunt impedance for signals of frequencies around 1.25 MHz below the picture carrier. Resistor R_1 terminates the high-pass m-derived filter section for minimizing reflections in the high-pass section.

Instead of using two-wire lines, the filter sections for the vestigial filter are made up of coaxial-line stubs. Coaxial lines have the same characteristics as two-wire lines regarding quarter-wavelength resonance and reactive quantities above and below resonance. The only difference is that the coaxial sections are inherently unbalanced lines, as discussed earlier, in which the outer conductor (shield) is at ground potential.

As shown in Fig. 10-14, coaxial-line stubs can be used with two-wire

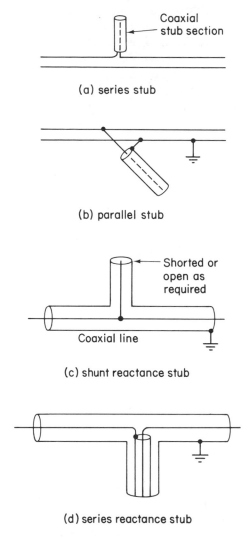

(a) series stub

(b) parallel stub

(c) shunt reactance stub

(d) series reactance stub

figure 10-14 coaxial-line stubs

lines, or with coaxial transmission lines for forming resonant or reactive sections. In Fig. 10-14a the coaxial stub is in series with one wire of a two-wire line, and if the stub is a quarter-wavelength long and open, it acts as a series-resonant circuit. (The transmission line must always be considered as the signal source; hence, it is the generator with respect to a stub, whether the latter is in series or parallel to the line.)

In Fig. 10-14b a coaxial stub parallels a two-wire line, and the length of the stub (and whether open or closed) again determines its characteristic.

With coaxial stubs the closed type is preferred to minimize losses at the open end. In Fig. 10-14c both the stub and line are coaxial types, with the stub section parallel to the line. Thus, the stub could be used to form a shunt reactance or a shunt resonant circuit. In Fig. 10-14d a coaxial stub is in series with the inner conductor of a coaxial transmission line. Note the use of the double-outer-conductor section, with both at ground potential.

The diplexer unit used to combine the FM and TV transmissions was shown in Fig. 8-11. The coaxial-cable equivalent is shown in Fig. 10-15

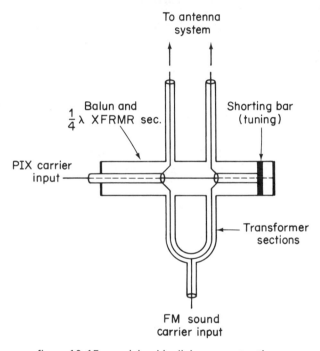

figure 10-15 coaxial-cable diplexer construction

with coaxial stubs forming balun section and transformer. A shorting bar is used for adjusting the length of the section at the right for tuning the system to the proper frequency section. Since high RF power is involved, the coaxial sections resemble metal pipes with a diameter of one or more inches as needed.

As shown in Fig. 10-16, two-wire line or coaxial sections form resonant circuitry in UHF oscillators. Similarly, such sections are used for resonant circuits in RF amplifiers. For the UHF oscillator (Fig. 10-16a) the collector is connected to one wire of the parallel-resonant line, while the other wire of the line couples to the base-input circuit. This oscillator can be compared to the Hartley type discussed in Chapter 3 and illustrated in Fig. 3-9, though it

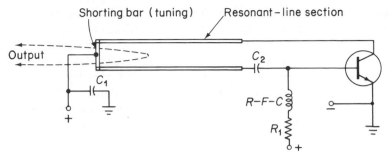

(a) ultra-audion type, resonant line section

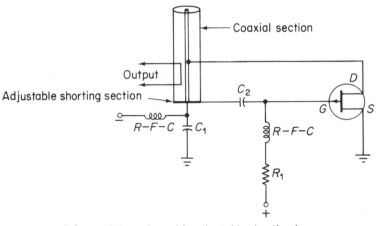

(b) coaxial section with adjustable shorting bar

figure 10-16 UHF oscillators, using line-resonant sections

is sometimes called an *ultra-audion* type. The collector line section can be considered as the collector inductance, while the resonant-line section at the base forms the base inductance. The reverse bias for the collector (positive for an *npn* transistor) is applied to the movable shorting bar (which tunes to resonance). Capacitor C_1 places the bar at *signal* ground, making it the same potential as the emitter. Thus, the emitter taps the inductance in similar fashion to the Hartley oscillator.

The output from the oscillator is tapped by using a single-loop inductance, as shown by the broken-line section in Fig. 10-16. Such a single-loop inductance is often termed a *hairpin loop*, and the closeness (degree) of coupling determines the amount of signal energy transferred, the amount of coupled load applied to the oscillator, and the resultant circuit Q.

For the oscillator of Fig. 10-16b, a coaxial-cable section is used for resonant-circuit purposes, and an FET unit replaces the transistor for illustrating the variations possible in such circuitry. Function is virtually the same as for the oscillator of Fig. 10-16a. The inner conductor of the coaxial cable is used for the drain element of the FET (or collector of a transistor). The outer conductor is not at direct (dc) ground, but is placed at signal ground by capacitor C_1. Output RF is obtained from a hairpin-loop system inserted into the coaxial element through small holes in the outer conductor, as shown. For both oscillators, RF chokes prevent signal-energy loss to ground or the power-supply sections, and capacitors C_2 block the dc potentials of the input circuits from that applied to the output.

The adjustable shorting bar often consists of a metal washer-type assembly mounted on a threaded rod connected to the tuning knob. RF energy is confined to the inside of the coaxial-cable section and does not penetrate through or leak to the outside of the tuning section. Because of the peculiarity of the phenomenon known as *skin effect*, RF signal energy flows on the outside of the inner conductor of a coaxial cable, and on the inside of the outer conductor.

This characteristic of RF signal-energy conduction can be more readily understood by reference to Fig. 10-17, which shows a magnified cross section

Conductor surface (skin)

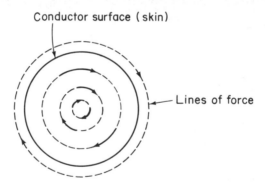

Lines of force

figure 10-17 skin effect

of a single wire carrying electric current. Current flow creates a series of magnetic line, and such fields are distributed not only within the wire core, but around the outside perimeter, as shown.

These magnetic-field lines represent the inductive factor along the length of the wire, which thus has reactive characteristics and opposes RF current flow. Since inductive reactance rises with higher-frequency signals, the opposition to alternating current within the wire rises as the frequency of the signal is increased. Consequently, currents for high-frequency signals find it difficult to flow through the inside of the wire, as normally occurs for very low frequency signals. Since, however, the first magnetic line on the outside of the wire is spaced a short (but definite) distance from the surface, the current flow established by the pressure of the voltage tends to flow on the outside of the wire where it meets with less opposition.

This "surface path" for the current at high frequencies is likened to the skin of the wire, and, hence, is called *skin effect*. Thus, to provide for the least opposition, wire diameters are increased to provide for more surface (skin) area at high and ultrahigh frequencies. In transmitting, copper tubing is often used, since the inner copper solids of the wire are no longer useful for current-carrying purposes and can be dispensed with. For receiver applications it is more practical to use heavy copper wire (size 18 or larger) for the coils, particularly since voltage and current levels are very much lower.

When coil diameters are increased to minimize skin effect, there is a corresponding rise in the distributed capacitance of the inductor. The larger-diameter turns thus increase the shunting effect of the capacitance between turns and can also affect the true resonance of tuning circuitry. Thus, the coils are usually designed with adequate spacing between turns to minimize the increase in capacitance effects. Fortunately, at higher-frequency operation, coil turns are less and, hence, do not become too large despite the larger-diameter wire and spacings.

Eventually, still higher frequency operation entails use of the line sections for resonant circuitry. If still higher frequency operation is necessary, the special microwave gear described in Chapter 11 is employed.

10-14. measurement, matching, and delay

When a section of transmission line is coupled to the resonant circuit of an RF amplifier or oscillator, as shown in Fig. 10-18a, wavelength measurements can be made. Since an open or closed line does not consume the signal energy picked up by the coupling loop, standing waves occur, as previously mentioned. Thus, an RF indicating meter can be used to seek out standing-wave loops for measurement purposes. Such a measurement line is often referred to as a *Lecher line*.

Instead of moving the indicating device along the line, a shorting bar can be used, as shown. At the shorting bar the standing wave of voltage will always be zero and the current standing wave will be at its maximum. Either the voltage or current loops can be used for measurement purposes by keeping the indicating meter stationary at the RF source end. Thus, as the shorting bar is moved, the loops of voltage and current change, and the distance the bar is moved between two loops (or two nodes) spans *one half-wavelength*.

When the distance between two loops is measured, the value so obtained can be inserted into either of the following equations for ascertaining the frequency of the signal generated by the RF oscillator (or amplified by the RF stage involved). For UHF, the measurement may be in inches; hence, we can

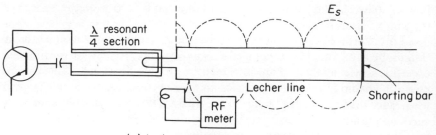

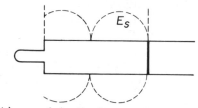

(a) lecher line measurement

(b) use of shorting bar

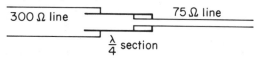

(c) use of closed line

figure 10-18 Lecher-line measurements and Z-matching

use the basic formula given in Chapter 1 [Eq. (1-7)], using $2d$ for distance in inches between two loops, for obtaining the frequency in megaherz.

$$\text{frequency (MHz)} = \frac{11{,}808}{2d} \qquad (10\text{-}28)$$

Thus, if we measure 24.5 in. between two voltage loops, we would ascertain the frequency as

$$\frac{11{,}808}{59} = 200 \text{ MHz}$$

If the distance is measured in meters, we can use Eq. (1-9) with $2d_m$ representing the distance between two loops in meters:

$$\text{frequency (kHz)} = \frac{300{,}000}{2d_m} \qquad (10\text{-}29)$$

For other distance values (feet, etc.) other formulas given in Chapter 1 can be modified accordingly.

As shown in Fig. 10-18c, a closed-line section can be used for impedance matching. Here, a quarter-wavelength section of Lecher line acts as a step-down transformer because the impedance is high at the open end and zero at the closed end. Thus, by tapping off certain points, a high-impedance line can be matched to a low-impedance one, as shown. The Lecher line can be reversed for impedance step-up purposes. An impedance match is obtained when a maximum amount of signal energy is transferred between generator and load, and standing waves are at a minimum.

As discussed in Chapter 8 and illustrated in Fig. 8-7c, sections of series inductance and shunt capacitance can be used for signal-delay purposes by tapping off at appropriate intervals and terminating the line in a load resistor to reduce line reflections. (See also Fig. 10-9b.)

The time in sections for signal energy to travel a section of line is found by

$$t = \sqrt{LC} \qquad (10\text{-}30)$$

Thus, if we wish to find the delay for the example of section specifications given in Section 10-10 for 75 ft, we obtain

$$t = \sqrt{0.36 \times 10^{-3} \times 10^{-9}}$$
$$= 0.6 \times 10^{-6} \text{ s } (0.6 \ \mu\text{s})$$

For multiples of the section calculated, this 0.6×10^{-6} value is multiplied by the number of sections involved. Thus, for two sections (150 ft) we obtain $2 \times 0.6 \ \mu\text{s} = 1.2 \ \mu\text{s}$.

questions and problems

10-1. Briefly define the term *half-section* as it relates to filter circuits.

10-2. Describe the basic functions of a π-type low-pass filter.

10-3. Explain briefly what factors contribute to the characteristic impedance of a filter section, and why this value is constant for added successive sections.

10-4. Define the term *constant k* as it relates to filter circuitry.

10-5. For a constant-k low-pass filter, one section has an inductance of 0.81 mH and a capacitance of 1000 pF. What is the cutoff frequency (f_C)?

10-6. What is meant by the term *m derived*?

10-7. How does the balanced filter differ from the unbalanced type? Explain briefly.

10-8. Explain briefly in what manner a bandstop filter allows all signals to pass through it except those clustered around a specific frequency.

10-9. What advantages and disadvantages does the two-wire line have over the coaxial cable?

10-10. What factors determine whether or not a transmission line is an *untuned* type?

10-11. Briefly define the terms *standing waves, loops,* and *nodes.*

10-12. Under what conditions is the impedance of a transmission line no longer constant along its length?

10-13. Briefly explain how the standing-wave ratio can be reduced in a transmission line.

10-14. What length of line must be used to form a parallel-resonant circuit, and specify whether or not the line must have a closed end?

10-15. When using a stub less than a quarter-wavelength (and closed at the end), what reactive component characteristic is obtained? Explain the reason for this.

10-16. Define the term *hairpin loop* and explain its purpose.

10-17. What is meant by *skin effect* and in what manner does this influence circuit functions at the higher frequencies?

10-18. In a Lecher line used for frequency measurement, the distance between two voltage nodes is 1 ft. What signal frequency is indicated?

10-19. The distance between two voltage loops in a Lecher line is 0.5 m. What is the wavelength of the signal (in megahertz)?

10-20. Show, by a simple drawing, how a closed line can be used as a step-up transformer for changing the input impedance of 75 Ω to an output impedance of 100 Ω.

10-21. How long does it take a signal to travel down a two-wire open line if the inductance value is 0.64 mH and the capacitance value is 1000 pF?

10-22. What is the Z_o of the line specified in Problem 10-21?

11
microwave principles

11-1. introduction

The microwave span of frequencies has not been standardized in terms of lower and higher limits, as has been done with VHF and UHF covered in Section 1.3. (See also Appendix J.) Generally, however, the microwave region is considered to extend from the upper regions of the UHF band (starting around 1000 MHz) through the SHF and EHF sections.

Microwave principles are an extension of the UHF factors covered in Chapter 10 for transmission lines, including standing-wave relationships, resonant sections, reactive components for forming filters, plus fields established by voltage and current applications. Because of the much higher signal frequencies involved, however, there is even a greater deviation from the common forms of transmission lines, resonant and reactive circuitry, and signal-generating systems encountered at the lower frequencies. (Antenna principles and propagation factors relating to UHF and microwave regions are covered in Chapter 12.)

Microwaves are extensively used for information-relay systems in communications, for line-of-sight transmission, and other services, including radar. The latter, however, is not strictly a communication system, but a sensing device for locating and tracking planes, missiles, satellites, etc.

11-2. waveguide dimensions

For the transfer of microwave signals between generators, amplifiers, antennas, etc., *waveguides* replace the two-wire lines and coaxial cables. The waveguides are so named because they guide the signal waveforms through their tube-like or pipe-like enclosures in a fashion that begins to simulate the propagation characteristics in free space. Also, because of skin effect, the microwave signal energy is confined to the inside of the metal waveguides and radiation losses are kept at a minimum.

Waveguide sizes determine the range of signal frequencies that can be accommodated. For the waveguide dimensions there is a relationship to the two-wire transmission line, as shown in Fig. 11-1a. Assume that quarter-

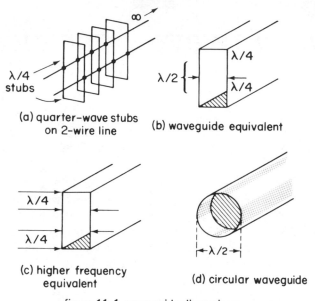

(a) quarter-wave stubs on 2-wire line

(b) waveguide equivalent

(c) higher frequency equivalent

(d) circular waveguide

figure 11-1 waveguide dimensions

wave stubs are attached to the top and bottom of a two-wire line and that such stubs are infinite in number and close enough to each other to touch. In such an instance the tubular section shown in Fig. 11-1b is formed, since the infinite number of quarter-wave sections form the solid walls. Thus, the wide section of the waveguide now is one half-wavelength for the lowest-frequency signal that will be propagated through it.

For higher-frequency signals the equivalent condition shown in Fig. 11-1c will occur. Here, the higher-frequency signals occupy only portions of the upper and lower sections of the waveguide. It is as though the center portions on each side represent a two-wire line that has widened as the frequency of the input signals has been raised.

The narrow dimension of the guide determines the maximum signal potentials that can be used. Widths may be as low as a quarter-wavelength, for the cutoff frequency, and up to a half-wavelength (in which case all side dimensions are the same and form a square). The manner in which the signal is injected into the waveguide determines certain propagation characteristics, as detailed more fully later.

A circular waveguide is also used (Fig. 11-1d). The diameter is one half-wavelength for the lowest (cutoff) frequency. Generally, both the rectangular and circular waveguides are designed for a wavelength two or three tenths above the cutoff frequency to provide for the necessary safety margin that assures reliable operation.

As shown in Fig. 11-2, waveguides can be perfectly straight, or bent and

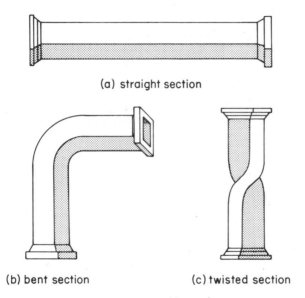

(a) straight section

(b) bent section   (c) twisted section

figure 11-2 waveguide sections

twisted as required. Bends are necessary to route the waveguide around corners, though the curve must be as gradual as possible to avoid reflections within the guide and resultant standing waves. The bending can be along either the narrow or wide dimension. On occasion flexible guides (made of pliable plastics with a metallic inside coating of chromium or other conducting surface) are used when more than simple bends are involved.

When the polarity of the signal must be changed, the waveguide is twisted as shown in Fig. 11-2c. Field rotation is necessary for proper polarity when applied to antenna systems (parabolic reflectors and flared waveguides as more fully described in chapter 12).

When signal energy travels through the hollow within a waveguide, it has the same characteristics as a propagated wave in free space and consists of electromagnetic and electrostatic fields generated by the signal currents and voltages. In free space, however, waves tend to spread, while within the waveguide the walls confine the fields to the boundary limits set by the waveguide configuration.

Depending on the manner in which signal energy is injected into the waveguide, magnetic and electric fields are set up in certain fixed positions within the guide. Thus, for a particular mode of operation, the magnetic fields may be positioned as shown in Fig. 11-3a. Here the field intensity is

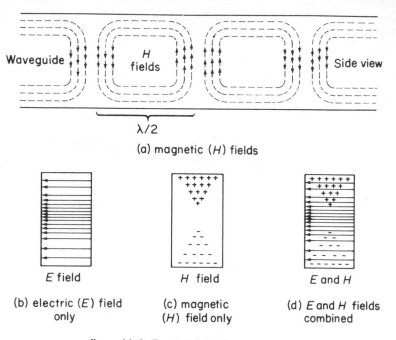

(a) magnetic (H) fields

E field   H field   E and H

(b) electric (E) field   (c) magnetic   (d) E and H fields
        only   (H) field only   combined

figure 11-3 E and H fields in waveguide

highest at the inside surfaces of the waveguide to correspond to similar high current conditions. Field strength tapers downward toward the center of the guide for each half-wave section, as shown. Instantaneous values are shown, for the polarities change as the cycle of the signal energy changes. The magnetic field is designated as the H field.

The accompanying electrostatic fields making up the complete signal-energy wavefront are designated as the E field, and the lines are at right angles to the H fields. For the particular mode of operation shown in Fig. 11-3a, the E fields would appear as in Fig. 11-3b when viewed from the waveguide end. Since the arrows are shown closer toward the center, this represents the high intensity of the field along the half-wavelength dimension.

The H fields are shown in cross-section representation in Fig. 11-3c as viewed from the waveguide end. The polarity represents an instantaneous condition. The combined E and H waves are shown in Fig. 11-3d and represent a particular operating *mode*. Such modes are identified as transverse magnetic (*TM*) or transverse electric (*TE*). For the *TE* mode, the electric-field lines are perpendicular to the waveguide length; hence, no *E* lines are parallel to the direction of the propagation within the guide. For the *TM* mode, the *H* field is perpendicular to the waveguide length.

Subscripts are also used to identify the mode characteristics, and for Fig. 11-3d, the mode is $TE_{0,1}$ for this rectangular waveguide. The first subscript number shows how many half-wave patterns of transverse lines occur for the shorter dimension of the waveguide. Since the *E* fields are transverse for this example, there are no amplitude changes through the center of the cross section of the short dimension; hence, the first subscript is 0. The second subscript relates to how many transverse half-wave patterns are found through the long dimension (cross-section center). Since the *E* field changes from zero to maximum and again to zero (as shown in Fig. 11-3b), the second numerical subscript is 1.

For circular waveguides, the initial numerical subscript indicates how many *full-wave* transverse fields are encountered once around the circumference. Thus, for Fig. 11-4a, where a *TM* mode is indicated, there is no change of amplitude of the fields around the circumference, indicating a 0 subscript. For the circular guides the second subscript shows how many *half-wave* field patterns occur through the diameter. Thus, for Fig. 11-4a the minimum-maximum-minimum changes equal a half-wavelength; hence, the second subscript is 1.

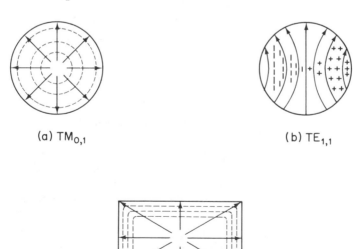

(a) $TM_{0,1}$ (b) $TE_{1,1}$

(c) $TM_{1,1}$

figure 11-4 TM and TE waveguide modes

A $TE_{1,1}$ mode for a circular guide is shown in Fig. 11-4b. Since a TE mode is indicated, the E transverse lines show a change around the circumference equal to a full wavelength (a rise, fall, rise, and fall of amplitude). A representative TM mode for the rectangular waveguide is shown in Fig. 11-4c.

Energy is injected into (or obtained from) a waveguide by using either an electrostatic or an electromagnetic probe, as shown in Fig. 11-5a, where an

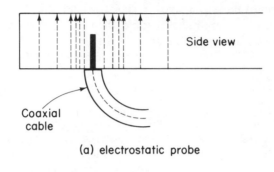

(a) electrostatic probe

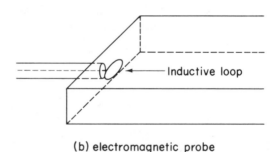

(b) electromagnetic probe

figure 11-5 waveguide excitation methods

extension of the inner cable of a coaxial line is placed in the center of the wide portion, one quarter-wavelength from the closed end of the waveguide. Thus, the probe parallels the narrow dimension but does not span it completely. The amount of energy injected is influenced by the probe position and the degree to which it extends into the waveguide. When signal energy is present in the probe, the current flow established electrostatic fields and in turn the E lines that are set up leave the probe and establish themselves as shown. The resultant magnetic fields (with the electric) thus comprise the wave energy, which propagates down the length of the guide.

Waveguide excitation with a magnetic-field probe is shown in Fig. 11-5b. Here the inner conductor of the coaxial cable is formed into a single loop and connected to the wall of the guide, as shown. When signal-energy is present within the cable, the high-value currents in the loop form strong magnetic fields, which expand and fill up the immediate area of the waveguide. If the frequency of the RF signal is above cutoff for the waveguide, the wavefront of energy will again travel down the length of the waveguide.

The magnetic coupling of Fig. 11-5b is applied at one of the places where the magnetic field is of highest intensity. For the waveguide shown, a probe could be placed into either adjacent waveguide wall since high current areas exist, as shown in Figs. 11-3 and 11-4. A similar loop arrangement can be used for removing signal energy from the waveguide.

11-3.   reactive sections

Reactive or resonant elements can be introduced into waveguides either by using certain partitions or by attaching additional waveguide sections having a quarter-wavelength (or less) dimension, as shown later. The reactive partitions are shown in Fig. 11-6 and consist of metal plates positioned with

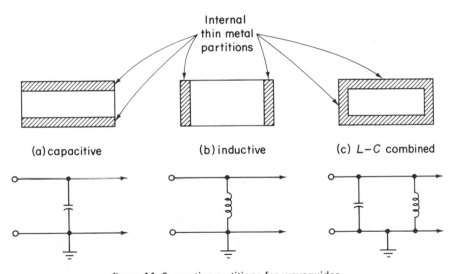

figure 11-6 reactive partitions for waveguides

respect to the E or H fields. In Fig. 11-6a the reactive partitions are added perpendicular to the E field and create a capacitive reactance.

To form an inductive reactance the plates are positioned in a vertical plane with respect to that of the H field, as shown in Fig. 11-6b. As with the reactive plates of Fig. 11-6a, a shunt reactance is formed having a value dependent on the width of the plates. When the two forms are combined (Fig. 11-6c), a parallel-resonant circuit is formed. Since reactances so formed are equal in value but opposite in polarity, the section shown is purely resistive. Since the effective cross area of the guide has been reduced, maximum voltage ratings are reduced accordingly.

As shown in Fig. 11-7, waveguide stubs can be used to form reactive sec-

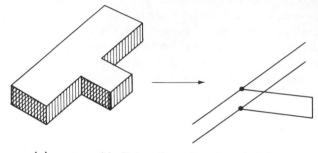

(a) narrow–side, T–junction, closed–end stub

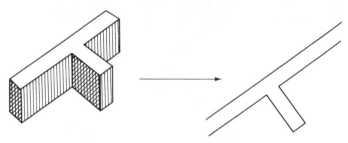

(b) wide–side, T–junction, closed–end stub

figure 11-7 T-junctions

tions. Though open- as well as closed-end stubs can be employed, the closed-end types are favored, since there is no radiation loss of signal energy. The attached sections are termed *T junctions* and may be applied to one of the narrow sides, as in Fig. 11-7a, or to a wide side, as in Fig. 11-7b.

If the waveguide of Fig. 11-7a is operated in the $TE_{0,1}$ mode and the T junction is in the plane of the H field, the reactive section is termed an H-type junction. Such an H-type junction is in parallel with the equivalent transmission-line characteristics of the waveguide, as shown at the right. At a quarter-wavelength a parallel-resonant circuit is formed, as was the case with two-wire stubs. For other reactive characteristics, the T junction is made less or more than a quarter-wavelength.

With the T junction lying in the plane of the E fields, a T-type junction is formed, as in Fig. 11-7b, for the same $TE_{0,1}$ mode. Now the equivalent system is as shown at the right with the junction in series with one side of the equivalent transmission line.

Thus, the T junctions can behave as a parallel reactance (or resonance) as well as a series type, and high-pass, low-pass, bandpass, and bandstop filters can be formed, just as was the case with the two-wire lines discussed in Chapter 10. Several T junctions can be used simultaneously to provide for the necessary shunt and series circuit elements.

11-4. *cavity resonators*

If a section of waveguide is totally enclosed and is one half-wavelength square, as shown in Fig. 11-8a, a parallel-resonant circuit is formed having

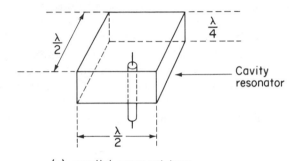

(a) parallel–resonant type

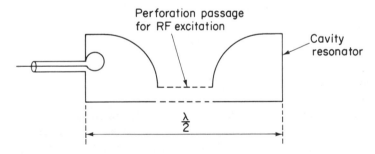

(b) magnetic-loop type with perforation

figure 11-8 cavity resonators

a high Q and, hence, excellent circuit efficiency. Thus, at microwave frequencies such cavity resonators are extensively used to replace the physical capacitors and inductors, which could no longer provide the resonant characteristics needed.

As with waveguides, cavity resonators could also operate at frequencies much lower than normally used, but their size would be prohibitive. At a couple of megahertz, for instance, the cavity resonator would require a width of approximately 150 ft. Thus, from the practical standpoint, applications start at around 3 gigahertz (GHz) or higher, where widths are in inches or fractions of inches.

Instead of the rectangle of Fig. 11-8a, a circular metal container can slso be used for a cavity resonator, provided the one half-wavelength is applied to the diameter. As shown, energy can be applied to (or taken from) the cavity with the electrostatic probe described for the waveguide by placing it at the center of the wide area (and thus one quarter-wavelength from the sides).

A magnetic loop-type probe can also be used, as shown in Fig. 11-8b. Another method that can be used for exciting the waveguide with RF energy is to pass the signal energy through perforations in the upper and lower surfaces. As shown, the top and bottom sections are spaced more closely together and holes included for accommodating an electron stream of signal energy. When such an electron stream is directed through the perforations, it influences the cavity area and sets up fields. Thus, an electron stream approaching the perforations repels free electrons in the structure, creating some current flow. An increase in electron activity thus provides signal injection. Once signal energy is present within the cavity, there is an energy interchange between the capacitance and inductance factors of the resonator, just as with resonant coil-capacitor combinations. The electron-stream principle is used primarily in resonators that are a self-contained part of the special tubes used in the microwave regions, as described next.

Tuning for the cavity resonators can be done by compressing the metal sides to alter the internal dimensions. Other methods include the use of an adjustable metal plug or disk for making slight alterations to the reactive components, thus changing the resonant frequency slightly.

11-5. the klystron tube

The *klystron* is a special tube for use at microwave frequencies. It features one or more self-contained cavity resonators; hence, it has the resonant circuit built into its basic structure. Klystrons are available in various sizes and serve as amplifiers (two or more cavities) and as single-cavity reflex-type oscillators. A typical type is shown in Fig. 11-9.

The perforated-type cavity resonator shown in Fig. 11-8b is used with the cavity holes acting as grid structures through which the electron stream is directed. Electrons approaching and leaving these grids set up a mode of oscillation determined by grid spacings and cavity dimensions. The anode is made negative in polarity and thus repels the electrons stream back toward the grid structure to sustain oscillations. During the initial pass through the grids, the electron stream is influenced by the changing polarity of the cavity

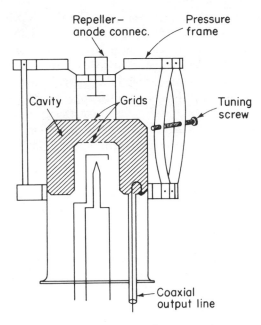

figure 11-9 Klystron microwave tube

oscillations. Thus, on their return path they reinforce the oscillations. The repeller-plate voltage can be varied according to a modulating frequency, thus producing a frequency-modulated output signal from this oscillator. As shown, the output is obtained by using a coaxial-cable pickup loop.

As shown, a pressure frame is mounted directly on the tube and is adjustable by the tuning screw. The latter separates or closes the two spring-like posts, which in turn change the pressure on the cavity and alter the spacing between the grid structures. This change alters by several megahertz the frequency of the signals generated.

In the dual-cavity klystron, amplifier signals to be processed are entered into one cavity, and the electron stream that goes through the grids is velocity modulated. In turn, the electron stream imparts its energy to activate a second cavity from which the amplified signal is obtained. The electron stream is attracted by a conventional anode with a positive polarity.

11-6. the magnetron

The *magnetron* is another widely used microwave oscillator for use in regions around 30 GHz. Essentially, the magnetron is a diode, since it only contains the filament-cathode structure and an anode. The latter, however, has a number of cavity resonators within its structure and, hence, has self-contained resonant circuits, as was the case with the klystrons. For the magnetron, the anode may be formed from solid copper for heat dissipation purposes, particularly when output power levels run into the megawatt ranges.

The individual cavities are coupled to each other through the capacitances at the common openings that face the central area containing the cathode. The cavities may also be coupled together by using linking metal straps between cavities. Thus, the microwave energy developed within the individual cavity resonators is transferred around the circle of resonators and finally picked up by the magnetic loop formed by the single-turn coaxial-cable conductor, as shown in Fig. 11-10.

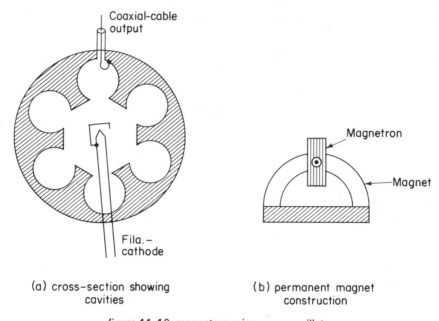

(a) cross-section showing cavities

(b) permanent magnet construction

figure 11-10 magnetron microwave oscillator

In Fig. 11-10b, a strong permanent magnet is coupled to the magnetron so that the magnetic fields thus created influence the electron flow within the magnetron structure. Thus, the electron stream is attracted by the positive-polarity anode and, at the same time, right-angle electron deflection occurs because of the magnetic fields. Hence, the electron stream assumes an elliptical path toward the anode and imparts energy to the cavities in passing.

Since cavity sizes are fixed, frequency alterations must be made by changing the anode voltage. Some magnetrons have been developed that provide for slug-screw adjustments to the cavity for frequency changes.

questions and problems

11-1. What factor contains microwave signals within a waveguide and prevents radiation losses?

11-2. How wide must the wide section be for operation of a waveguide near its cutoff frequency?

11-3. What precaution must be used when routing a waveguide around corners?

11-4. What is the purpose for giving a waveguide a single twist?

11-5. Define the terms E field and H field.

11-6. Explain what factors determine whether an operating mode for a waveguide is TE or TM.

11-7. Explain briefly the significance of the numerical subscripts used with the TE and TM designations.

11-8. Describe two methods for injecting and removing signal energy from a waveguide.

11-9. Briefly describe how reactive partitions are used to introduce capacitance and/or inductance into the waveguide structure.

11-10. Explain how T junctions can be used to form reactances and resonant circuits.

11-11. Why is it preferable to use closed ends in the T sections?

11-12. In what manner is the resonant frequency determined for (a) a rectangular waveguide? (b) a circular waveguide?

11-13. Describe three methods for injecting signal energy into a cavity resonator.

11-14. Explain briefly what methods can be used for tuning a cavity resonator.

11-15. Outline the process by which a reflex klystron generates a microwave signal.

11-16. Explain briefly how, in a klystron, the frequency of the generated signal is changed.

11-17. What is the purpose for using an external magnet with the magnetron?

11-18. How is the signal frequency of the magnetron changed? Explain briefly.

11-19. How is the energy of each cavity resonator of a magnetron added to adjacent cavities?

11-20. What type of path does the electron stream take within a magnetron, and what causes such a path?

12

antenna systems

12-1. introduction

Antennas serve as either the propagating load resistance of the transmitter or as the signal intercepter for the receiver; hence, they have a dual relationship known as *antenna reciprocity*. The sending and receiving functions of an antenna utilize identical characteristics relating to wavelengths, fields, and impedances. Hence, in many low-power applications, particularly in portable transmitter–receiver combinations (transceivers), the same antenna is used for both transmitting and receiving functions.

The maximum signal power transferred from the transmitter to the antenna (or from antenna to receiver) depends on the impedance-matching factors relating to transmission lines. Hence, much of the material in Chapters 10 and 11 applies to the discussions here.

As a transmitting device, the antenna must convert the signal components of voltage and current to a composite electric and magnetic field structure that will propagate through space. Conversely, during reception, the antenna must intercept the electric and magnetic fields making up the transmitted signal energy and reconvert it into equivalent values of voltage and current for amplification and demodulation purposes. The fundamental factors applying to these relationships are covered in this chapter and apply to antennas of all types.

12-2. antenna patterns and fields

In Fig. 10-10e, a quarter-wavelength open section of line was shown. If such a section of line is opened, the opposite-polarity fields of the two wires no longer cancel to reduce radiation losses, and an antenna structure is form-

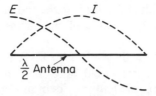

(a) half-wave dipole antenna

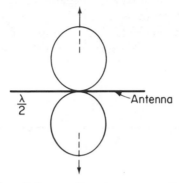

(b) radiation pattern in cross-section

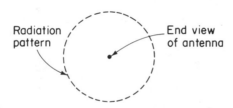

(c) radiation pattern, end view

figure 12-1 patterns of half-wave antenna

ed, as shown in Fig. 12-1a. (The generator has been omitted to simplify the illustration, though signal-feed methods will be covered later in this chapter.)

Since the line section was a quarter-wavelength long, the opened section now spans one half-wavelength and the voltage and current distribution is retained, as shown. Thus, each end still has a voltage maximum, with current reaching its peak value at the center. Such an antenna span is the shortest length possible for a straight element representative of a resonant circuit for the signal frequency to be handled.

Such a simple antenna is often opened at the center for attaching the transmission line, in which case it is known as a *dipole* antenna, or *Hertz* antenna, after Heinrich R. Hertz, the physicist and researcher. The half-wave antenna transmits or receives signals at right angles to its length, as shown in Fig. 12-1b. The individual sections of the radiation pattern are known as *lobes* and show the relative intensity of signal transmission or reception for various directions.

The directional pattern shown in Fig. 12-1b represents a cross-sectional view, since the same pattern would be evident if observed from any side parallel to the antenna length. This is evident in Fig. 12-1c, where the pattern is viewed from the antenna end and is shown as completely encircling the antenna. Thus, such an antenna is capable of operating for signals having directions at right angles to the antenna length.

When a transmitter feeds signal power into the antenna, electric and magnetic fields are built up around the antenna, as shown in Fig. 12-2. The

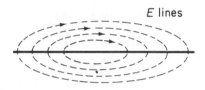

(a) electric field

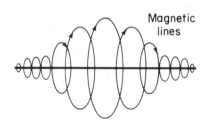

(b) magnetic field

figure 12-2 fields of half-wave antenna

E lines parallel the antenna length (Fig. 12-2a), while the *H* (magnetic) lines encircle the antenna, with highest amplitudes at the center (Fig. 12-2b). Since the signals are alternating current in nature, they build up with a specific polarity, collapse, and build up again with opposite polarity.

During the time the ac signal drops to zero between alternations, the fields would tend to collapse into the antenna. Because of the rapid change of polarity for high-frequency signals, however, new fields of opposite polarity emerge from the antenna before the existing ones can collapse back into the antenna. Hence, fields are forced beyond the influence of the antenna structure and propagate into space at the speed of light.

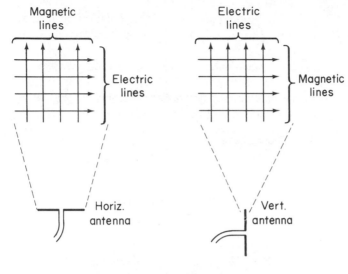

(a) horizontal polarization   (b) vertical polarization

figure 12-3 antenna polarization

The composite fields that make up the signal in free space consist of expanding magnetic and electric lines, as shown in Fig. 12-3. When the receiving or transmitting antenna is in a horizontal position with respect to earth, the propagated wave is *horizontally polarized*, with the electric lines also in a horizontal plane. When the antenna is in a vertical position, as shown in Fig. 12-3b, the wavefront of signal energy now has the electric lines in a vertical position, a situation known as *vertical polarization*. For a maximum transfer of signal energy between transmitting and receiving antenna, one of the essential factors is the maintenance of identical polarization between the two antennas. Other factors consist of antenna height, proper positioning with respect to the radiation pattern, angle of radiation, and atmospheric conditions.

Most television signals are horizontally polarized, though at the receiver some vertical polarization may be present in the signals because of reflections and inversions during propagation between antennas. When signals are transmitted or received at ground levels, the horizontally polarized antenna structure suffers shunt reactive losses because of the capacitance set up between the antenna and ground. Hence, vertical polarization is preferred.

When transmitting with vertically polarized antenna systems, such as standard AM, a single quarter-wavelength vertical antenna is used, as shown in Fig. 12-4. The bottom of the antenna is grounded, thus eliminating the need for a half-wavelength to achieve resonance for the frequency of the signal to be transmitted. When the vertical antenna is grounded, as in Fig. 12-4a, the system simulates the characteristics of a half-wave antenna because of

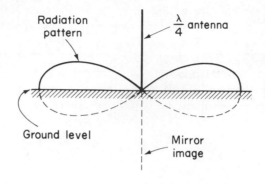

(a) mirror-image effect

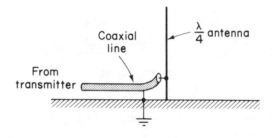

(b) signal feed

figure 12-4 vertical (grounded) Marconi
antenna

the *mirror-image* effect. This is as though the earth presented a mirrored extension of the quarter-wave section, as shown in Fig. 12-4a, and the equivalent half-wave section again has a figure-eight radiation pattern shown in Fig. 12-1b. Half, however, is the theoretical mirror image as shown. Thus, such an antenna propagates in all directions along a horizontal plane. Such an antenna is referred to as the *Marconi* type, after G. M. Marconi, the physicist who contributed much to early electronic knowledge.

Signal feed to the Marconi antenna can be performed as shown in Fig. 12-4b, using a coaxial cable for an unbalanced coupling (one end of the antenna grounded, and the outer conductor of the coaxial line also grounded). Since the impedance of the vertical antenna is zero at ground, the impedance rises along its height. Thus, the inner conductor of the coaxial line is attached at a place where an impedance match is obtained.

The term half-wavelength antenna means that the unit is used for a signal having a frequency corresponding to the antenna length. Hence, if lower-frequency signals are used, antenna efficiency and effectiveness drop sharply.

This is not to say such operation is impossible, since virtually any metallic object will pick up transmitted signals to a degree, depending on the strength of the signals in the area. Similarly, a horizontal receiving antenna could be used for vertically polarized waves if the arriving signals were sufficiently strong to permit normal demodulation and amplification. For low-power transmission or distant reception, however, factors regarding length, height, orientation (pointing the receptive angle of the antenna into the desired direction), and polarization are important.

Where multiple-frequency signals are involved (such as the antennas for television reception), the radiation pattern changes sharply for signals higher than those at the half-wavelength. If, for instance, a half-wave antenna receives signals of twice the fundamental frequency, the antenna may be considered to consist of two half-wave antennas of the higher-frequency signals joined together to form a single length, as shown in Fig. 12-5a. Thus, it now represents a full-wave antenna for the frequency now used.

In Fig. 12-5a, voltage and current distributions are now out of phase for each half-wavelength. In Fig. 12-5b, the individual halves of the antenna would have a figure-eight pattern, except for the influence of the respective

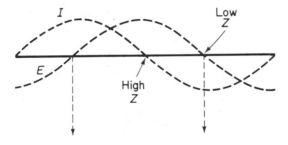

(a) simulated full–wave antenna

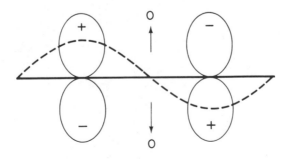

(b) radiation pattern

figure 12-5 full-wave antenna characteristics

polarities. As shown, out-of-phase conditions are present in both the true vertical and the true horizontal planes, and the 180° phase difference results in signal cancellation in these directions, both for a receiving antenna as well as the transmitting types.

At angles other than the horizontal or vertical with respect to the antenna position, only partial out-of-phase conditions prevail; hence, reception or transmission is possible in such angles, as shown in Fig. 12-6. This cloverleaf radiation pattern means that such an antenna has a pickup sensitivity from four separate directions, with each lobe at a 54° angle with respect to the antenna plane. Similarly, a transmitting antenna a full wavelength long would also exhibit such a cloverleaf transmission characteristic.

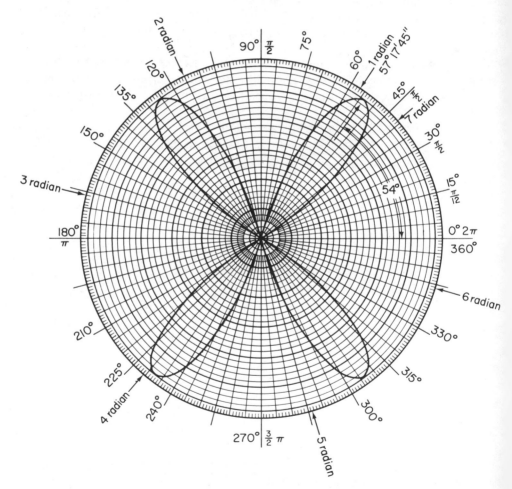

figure 12-6 full-wave antenna radiation pattern

If the same antenna is used for signals three times the fundamental, it behaves as an antenna one and one-half wavelengths long and the cloverleaf lobes again appear, though with a 42° angle. At higher frequencies, secondary lobes of low sensitivity also appear clustered around the junction of the major lobes. Thus, multiple-directive results are obtained. While this may have some advantages for signals arriving from various angles, it also increases interference problems.

Half-wavelengths for specific frequencies can be found by using the various equations given in Chapter 1. Equation (1-4) must, however, be modified for half-wavelength calculations by dividing the results by 2:

$$\frac{\lambda}{2} = \frac{15 \times 10^7}{f(\text{Hz})} \tag{12-1}$$

$$\frac{\lambda}{2} = \frac{150}{f(\text{MHz})} \tag{12-2}$$

Because of the effects of the wave traveling along a wire instead of free space, and also because of the *end effect*, which results from capacitance at the ends of the antenna, a correction factor is necessary, particularly for signals having frequencies above 30 MHz. Multiplication by 0.94 should be made for a close approximation to the required *electric length* rather than the physical length. Thus, for 60 MHz a half-wave antenna would have a length found as follows:

$$\frac{150 \times 0.94}{60} = 2.35 \text{ m}$$

The following are also useful:

$$\frac{\lambda}{2} \text{ (ft)} = \frac{492 \times 0.94}{f(\text{MHz})} = \frac{462.5}{f(\text{MHz})} \tag{12-3}$$

$$\frac{\lambda}{2} \text{ (in.)} = \frac{5550}{f(\text{MHz})} \tag{12-4}$$

Thus, using 60 MHz again for an example, we have

$$\frac{462.5}{60} = 7.7 \text{ ft}$$

For proof we can multiply the number of feet by 0.3048 to obtain meters: $7.7 \times 0.3048 = 2.3469$ m.

12-3. ionosphere factors

The distances over which satisfactory reception of transmitted signals is obtained vary considerably for the numerous broadcasting facilities. The wavefront of propagated energy levaing the transmitting antenna may diminish in intensity as it speeds toward distant receiving antennas because of ground absorption and expansion with distance. Also, losses occur because of atmospheric conditions, misoriented antennas, improper polarization, or because of objects obstructing the signal paths.

Some signals can be transmitted great distances around the earth because of the reflective characteristics of ionic layers that exist above the earth. Other signals penetrate such layers with a minimum of loss and, consequently, are used for satellite communication purposes and for maintaining audio and visual contact with traveling astronauts.

The ionic layers are termed the *ionosphere*, and for lower-frequency signals (such as the AM broadcast band) most of the signal energy is reflected back to earth. After being reflected to earth the signals can be reflected upward again, producing a skip effect and permitting reception over many miles. As the frequency of the transmitted signals increases, reflections occur from different layers, and by utilizing specific angles of radiation with respect to earth, predicted areas of reception are possible. After about 30 MHz, greater signal penetration of the ionosphere occurs, and because of the changing characteristics of the ionosphere, conditions of reflection may change often.

For VHF transmissions (including television), conditions of reflection become unreliable and most of the energy passes through the ionosphere. The variations in the ionic characteristics occur because of the nature of the ionosphere. The ultraviolet radiation reaching earth from the sun influences the upper atmosphere and creates layers of free electrons and ions. Such ionization levels increase during peak periods of sunspot activity (when ultraviolet radiation increases considerably).

The ionic layer nearest earth is designated as the *D* layer and ion intensity reaches its greatest peak around noon when the sun is highest. The *D* layer has a height of approximately 50 to 100 kilometers (km). The next highest layer is the *E*, which also exists only during daylight and ranges in height from about 90 to approximately 150 km. The *E* layer exhibits some erratic behaviour and the ionization levels may change appreciably each hour.

Since the *D* layer is lowest, more atmospheric particles exist and there is greater interactivity between electrons and ions, resulting in increased signal absorption. Similar signal absorption also occurs for the *E* layer, since particle activity is also high. Radio waves transmitted vertically from earth, however, are usually reflected back by the *E* layer to a greater extent than from the *D* layer.

The most unstable layer area exists directly above the *E* layer and has been referred to as the *sporadic E*. This layer may be present at any time during day or night in all areas above the earth, and results are usually unpredictable regarding the time and degree of signal reflection. The instability has been attributed to ionization by particle radiation from the sun in addition to the ultraviolet radiation.

The highest ionic layers contribute most to long-distance radio communications. During the night there exists a region termed the F layer, which ranges in height from 140 to about 240 km. During daylight the F layer splits into two separate ones, the F_1 and the F_2, with the latter extending to heights of 200 to about 350 km. Again ionization levels depend on the presence of the sun and reach a maximum around local noon time, declining slowly as the day wanes. Some ionization remains during the night, however, for the combined F_1 and F_2 layers, making the F (night) layer valuable for long-distance communications during evening hours also. The particle collision rate is less for the F layer; hence, there is less absorption and greater signal reflection.

The reflection characteristics of the ionic layers are shown in Fig. 12-7. Ionosphere reflection occurs when the amount of ionization present has a minimum of absorption, when a specific frequency is used, and when the proper angle is present. Thus, for signal S_1, penetration of the layers may occur and none is reflected back. For a change of angle such as that of S_2,

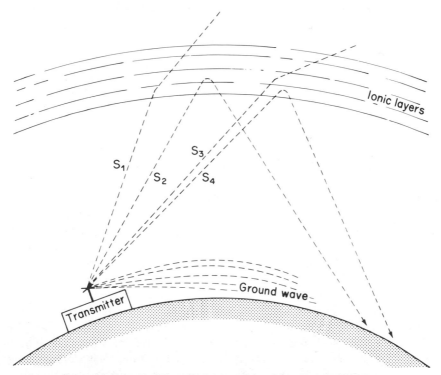

figure 12-7 ionosphere effect on angle and frequency of waves

however, the same signal, in terms of frequency, now reflects back to earth, as shown. Similarly, if a signal has a certain frequency, S_3, it may penetrate the ionosphere; but if the frequency is reduced, a point is reached where reflection again occurs, as for S_4. The frequency at which the division occurs for penetration and reflection (50 per cent of the signal penetrates and 50 per cent is reflected) is termed the *critical frequency*.

Another term in use is *maximum usable frequency* (MUF), which refers to the ability of the ionosphere to reflect transmitted waves between two points (such as the location of the transmitter for S_4 in Fig. 12-7 and the point where the reflected wave strikes earth on its return). Again, the MUF relates to the frequency when 50 per cent of the signal is reflected and a like amount penetrates the ionosphere. The National Bureau of Standards regularly publishes MUF predictions for the E and F layers as a guide for predicted reliability during certain time intervals and distances.

Commercial radio links often use the *troposcatter* method for distance communication. For both VHF and UHF there is a tendency for transmitted waves to undergo a scattering effect in the atmosphere, plus some tropospheric bending. The latter comes about because of the change of refractive index when there are temperature differences in the air-mass layers above earth. Thus, though VHF and UHF propagated waves are of a straight-line nature (sometimes called *line-of sight* transmission), special antenna networks for both transmitting and receiving make possible over-the-horizon reception for certain commercial services.

12-4. antenna couplings

Transmission lines can be coupled to antennas at the ends (high-impedance point) or at the center (low-impedance point) depending on requirements. For FM and TV reception the transmission lines are low-impedance types; hence, the coupling methods shown in Fig. 12-8 may be employed. In Fig. 12-8a the dipole is formed by opening the half-wave antenna rod, and the twin-lead-type transmission line is attached thereto.

Though the impedance at the opened point should be zero in theory, some capacitance is created between the open ends and some reactive effect occurs. Also, depending on antenna diameter, some skin-effect opposition may be present, and these factors contribute to a total impedance of approximately 75 Ω, as shown. Thus, if a 300-Ω transmission line is used, a mismatch will occur, resulting in a reduction of the maximum possible transfer of signals between the antenna and receiver.

The current loop in an antenna is used as the reference for determining what is termed the *radiation resistance* in either a transmitting or receiving antenna. In a transmitting system signal power is fed to an antenna, and volt-

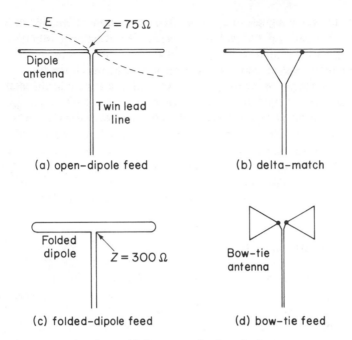

figure 12-8 antenna feed methods

age and current measurements would indicate that power is consumed. Yet, the antenna itself does not consume such power but only converts it to the type of energy that can be propagated. The equivalent resistance that consumes such power is, therefore, called radiation resistance.

For the receiving antenna the radiation resistance can be defined as the ratio of the power picked up by the antenna to the square of the effective current existing at the selected reference point on the antenna. Thus, in the dipole antenna the current-loop reference point makes the input impedance of the antenna equal to the radiation resistance. If the feed line were attached at the ends, however, the input impedance would no longer equal the radiation resistance, but rather would be determined by the ratio of E/I at the feed point.

Instead of forming a dipole, the half-wavelength can be left intact, as shown in Fig. 12-8b, and the wires from the transmission line fanned out and attached at points where an impedance match is obtained. This is possible because the impedance rises toward the right and left ends of the half-wave section, and spreading of the wires for a short distance is all that is required. This method is sometimes called a *delta match* because of its resemblance to that Greek letter.

A folded-dipole-type antenna can be used as shown in Fig. 12-8c. This double-rod antenna provides for a 300-Ω impedance at the opened section of the lower rod, thus forming a perfect match for the standard 300-Ω twin lead. (Most tuners of TV and FM receivers have balanced-line inputs for 300-Ω transmission lines, such as the twin-lead type, so a match is provided between the line and the 300-Ω input impedance of the tuner.)

The *bow-tie* antenna, sometimes used for UHF television, is shown in Fig. 12-8d. Again the line connects to the center where an approximate impedance match is obtained. Where a minimum of signal pickup by the line is intended, coaxial cables could also be used, with the outer (shield) conductor connected to one dipole rod and the inner conductor to the other. Since 75-Ω coaxial cables are common, an impedance match is readily obtained with the basic dipole-type antenna fed at the center. Losses, however, are greater for the coaxial cables used in television reception.

12-5. Yagi antenna

If an additional rod is placed in parallel to the antenna, but a short distance away, the new rod intercepts some signal energy and reradiates it, thus (in effect) reflecting it back to the antenna. Such a rod is termed a reflector and causes an increase in signal pickup (or transmission) for the direction perpendicular to the antenna side not having the reflector. The reflector rod is somewhat longer than the antenna, with dimensions more fully discussed later.

An addition of one or more rods is also possible ahead of the antenna structure, as shown in Fig. 12-9, to form an antenna known as the *Yagi*, after the noted inventor and physicist, H. P. Yagi. The Yagi construction not only increases gain and sensitivity but confines the radiation patter to only one direction, as shown by the dotted outline in Fig. 12-9. The sharpness of the lobe also minimizes reception of undesired signals or signal reflection from objects near the signal path.

As shown in Fig. 12-9, the forward rods are called directors and aid in directing some of the propagated signal energy toward the antenna for increased gain. The directors or reflectors are called *parasitic* elements to distinguish them from the antenna proper to which the transmission line is attached. The antenna section is called the *driven element*.

As also shown, all elements may be connected directly to the supporting metal rod, since their center areas are voltage-node points and have a minimum of impedance. The directors are often made progressively shorter for added sections to provide increased bandwidth. Increasing response to various frequencies is often necessary with the Yagi because its configuration increases Q, and if all elements are designed around a single frequency, the bandwidth becomes extremely narrow.

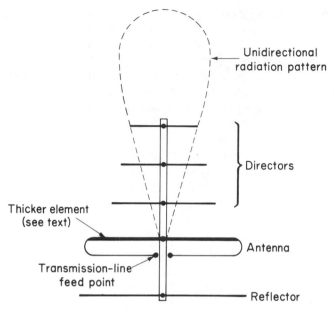

figure 12-9 Yagi antenna

For Yagi structures, the reflectors are usually made about 5 per cent longer than the antenna element, thus bringing the reflector to the approximate physical length rather than the electric length mentioned earlier for the antenna. Directors are made about 4 per cent shorter than the antenna for operation at or near the frequency for which the antenna is desired. For a broader band, however, successive directors should be progressively shorter, as shown. Spacings often are 0.15 wavelength between the antenna rod and the reflector, and 0.1 wavelength between the director and antenna, or between successive directors.

Gain increases rapidly for the Yagi as a reflector (or directors) is added. When a parasitic element is added to the driven element, the gain increases by about 5.5 dB over that of a single dipole (assuming a matched system). A four-element Yagi with proper design can achieve a gain of 9 dB over the single dipole. The addition of parasitic elements to the antenna system results in a decrease in the impedance at the feed point. The impedance can be increased again, however, by using a larger diameter unbroken rod of the folded dipole. For a four-element Yagi, a 300-Ω impedance is again achieved by making the diameter ratio between the unbroken rod and the broken rod (where the line is attached) four to one. Thus, if the unbroken rod element is 1-in.-diameter tubing, the section to which the line is attached should be $\frac{1}{4}$-in. tubing.

12-6. stacking factors

Antennas of any type can be stacked one above the other, using a single common transmission-line feed. When proper matching and phasing precautions are taken, the stacking of two or more antennas in the vertical plane increases signal sensitivity and also alters the radiation pattern over that obtained for the single dipole. With stacking signal sensitivity can be minimized in the vertical plane, for instance, and increased in the horizontal, thus increasing the direction-selectivity of the antenna. The disadvantage is the increased space occupied by the system, a factor that, however, becomes less bothersome as the signal frequencies are raised.

For obtaining a horizontal directivity pattern it is necessary to stack the antennas one half-wavelength apart, as shown in Fig. 12-10. Two rods or transmission-line sections are used, with the feed line connected at the center. As shown in Fig. 12-10b, however, the feed line can also be attached to the lower end of the stacked array, provided the section between the two antennas is transposed s shown.

With either of the methods shown, the phase relationships for signals are identical for each antenna, as shown in Fig. 12-10c. Thus, the in-phase

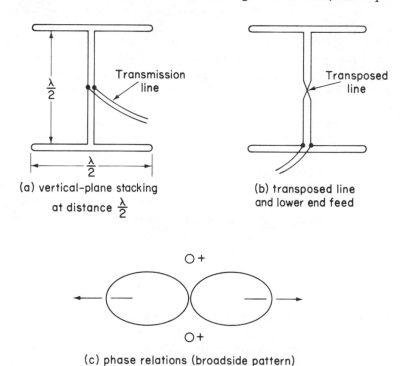

(a) vertical–plane stacking
at distance $\frac{\lambda}{2}$

(b) transposed line
and lower end feed

(c) phase relations (broadside pattern)

figure 12-10 antenna stacking

conditions prevailing along the horizontal plane increase the radiation pattern. There is a minimum sensitivity in the vertical plane because signal energy of a particular polarity alters for a half-wavelength. Thus, if the signal wavefront leaving the lower antenna is positive and travels upward, by the time it has spanned one half-wavelength the signal emanating from the upper antenna will have a negative polarity and cancellation will occur. Similarly, even though at any instant the signals for each antenna are in phase, when the energy of the upper antenna propagates toward the lower, the time interval is sufficient for the lower antenna's signal to undergo a polarity change, thus creating opposing signal conditions.

The stacking in essence parallels the impedances of the two antennas, resulting in a decreased total impedance presented to the transmission line. Also, since the spacings are based on one half-wavelength, a change of signal frequency alters the radiation characteristics of the array.

The type of pattern shown in Fig. 12-10c is called a *broadside* pattern because of its right-angle direction with respect to the stacked antennas. Additional antennas could be stacked in the vertical plane to increase the horizontal directivity of the array. With half-wavelength spacings between antennas, however, the bulk increases rapidly for lower-frequency systems.

If the interconnecting line sections of Fig. 12-10b were not transposed, the signal polarities at any instant would be opposite for the upper and lower antennas. Consequently, out-of-phase conditions exist for horizontal directions. For vertical directions, however, the signal traveling from one antenna to the other element spans a half-wavelength; hence, by that time the phase of the signals coincides. If, for instance, the signal were of negative polarity at the lower antenna and it propagated upward, its polarity would not change, but the signal leaving the upper antenna at the time the lower signal arrives will also be of negative polarity; hence, no cancellation occurs.

Thus, when no wire transposition is used for Fig. 12-10b, the radiation pattern of Fig. 12-10c would be in a vertical plane and would be called *end fire*, since it emanates from the ends of the stacked array. Again, more than two antennas could be used for increasing gain in a more narrow path.

12-7. reflectors and flared waveguides

In microwave practices a parabolic reflector is used with an antenna system for both transmitting and receiving signals in a straight-line path with a minimum of wavefront expansion during propagation. The result is a high degree of efficiency because of a minimum of signal loss and a pinpoint-type orientation highly suitable for direction-finding purposes.

Since microwave signals have frequencies approaching that of light waves, they behave in similar fashion in terms of reflection. Thus, a metallic parabolic reflector has the ability to direct microwaves in one direction just as a mirrored reflector is used for light-beam formation. For the microwave parabolic reflector, the antenna radiator must be placed at the focal point and the radiation from the antenna directed toward the bowl of the parabola, which then reflects the energy forward. As shown in Fig. 12-11a, a dipole ele-

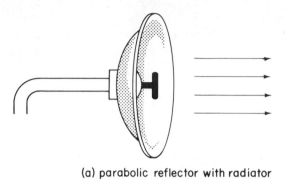

(a) parabolic reflector with radiator

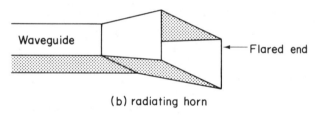

(b) radiating horn

figure 12-11 parabolic reflector-antenna and flared-end
waveguide

ment can be used, though often a metal shield is placed close to the antenna for aid in directing the signal energy toward the parabola.

As shown in Fig. 12-11b, the ends of a waveguide can be flared into a horn shape for propagating energy from it in a unidirectional path. Though such a flared horn-type structure directs the energy in one direction, dispersion does occur as the signal travel outward, and the narrow-path beam obtained from the parabolic structure is not attained with the flared-end radiator.

Rectangular metal shields can also be placed behind an antenna to act as a large-scale-reflector element for increasing forward gain for both transmission and reception. Sometimes right-angle sections are used (corner reflectors) for minimizing sensitivity toward the rear. As shown in Fig. 12-12, rods or wire mesh are used to decrease wind resistance. Such reflectors are placed at ground potential since they are parasitic elements, just as the

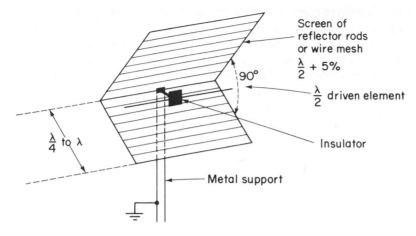

figure 12-12 corner reflector antenna system

reflectors and directors discussed earlier. As with the parabolic reflector, the corner reflectors are not the driven element, but must be used in conjunction with an antenna element.

12-8. field intensity

Stacked antenna arrays and multielement units such as the Yagi are often considered as having voltage or power gain. Actually, any antenna does not radiate all the power that is supplied by the transmitter and fed to it through the transmission line, since line losses and antenna efficiency are limiting factors. Complex antennas can, however, by directing propagated energy into specific directions, produce signal intensities at a given area that would be equivalent to an increase in power supplied to the antenna. This becomes evident by comparing the radiation pattern of the single antenna of Fig. 12-1c with those showing increased gain in specific directions, such as the Yagi in Fig. 12-9 and the stacked array in Fig. 12-10.

Thus, directive antennas can send the propagated wave in the direction of prescribed horizontal or vertical angles and increase signal strength in such desired directions by utilizing radiated energy normally transmitted at undesirable angles. This is done without an increase of the average power output of the antenna system. Hence, the reference to the *input power* to the final Class-C amplifier stage of a transmitter becomes less valid at VHF and UHF or microwave frequencies, since directive antennas are used. The signal delivered to a certain area may vary considerably in amplitude from that transmitted by another station, even though the two have identical input powers to their final RF amplifier.

Another factor in the VHF–UHF regions is antenna height above ground, where an increase raises the signal received in remote areas (line-of-sight transmission). Thus, the commonly used reference to transmitter power for TV and FM at the UHF–UHF areas is *effective radiated power*. This is the value obtained by taking the signal-power output of the transmitter's final RF stage, subtracting power loss in the transmission line, and multiplying the result by the power gain of the particular antenna system in use. Thus, a station might use 50 kW to the antenna, but because the antenna produces a gain of 3, would have an effective radiated power of 150 kW.

Still another consideration is the signal attenuation that tends to increase as the transmitted signal frequency is raised. Thus, where a lower-band (Channels 2 to 6) television station may obtain an effective-radiated-power rating of 100 kW from the FCC, stations in the Channels 7 to 13 range would obtain a rating allocation over 300 kW. For UHF stations, a still higher effective-radiated-power rating would be issued by the FCC, with powers to 5 megawatts (MW).

Since antenna height has a bearing, the FCC will limit antenna height to meet certain conditions. Thus, a 100-ft limit might be imposed for a certain effective radiated power, and if a higher antenna is to be used, the effective radiated power may have to be reduced accordingly. Specific restrictions depend on the geographical location of the station in terms of terrain, the number of other stations in the area, etc.

The field intensity or field strength at a given distance from the transmitter is usually specified in *microvolts per meter*. This term refers to the number of RF-signal microvolts intercepted by a length of wire exactly 1 m long, properly oriented for the signal being received and also polarized correctly with respect to the transmitting antenna. Thus, the microvolts-per-meter expression relates to the voltage obtained at the particular height at which the measurement is taken. Obviously, a higher reading can be obtained if the wire were raised (conceding factors such as horizontal polarization), but the intent is to specify the reading at a selected height.

Field gain refers to the amount of signal increase at a selected point some distance from the transmitter antenna for a change of input power, antenna directivity, antenna location, etc. Thus, if a simple transmitting-antenna system produces a reading of 250 μV on a field-strength meter, and a multi-element array increases this reading to 500 μV, there is a field gain of 2 for the larger unit. Since the values are expressed in terms of voltage, the decibel difference is 6. For field-power measurements (or for effective-radiated-power measurements at the transmitter), a doubling of power would indicate a 3-dB change ($10 \log P_1/P_2$) in contrast to voltage (or current) doubling, where a 6-dB change occurs ($20 \log V_1/V_2$).

questions and problems

12-1. Briefly define the term *antenna reciprocity*.

12-2. Where, on a Hertz antenna, are high and low impedance points found?

12-3. Briefly explain what is meant by a *horizontally polarized* antenna.

12-4. What factors make it preferable to use vertical polarization for antennas used near ground?

12-5. What is the *mirror-image* effect in relation to the Marconi antenna? Explain briefly.

12-6. What type of radiation pattern is obtained when an antenna is operated at twice the frequency for which it was originally designed? (Explain your answer.)

12-7. What is the *electric length* in meters of an antenna rod designed for operation at 100 MHz?

12-8. Briefly explain what effect the ionosphere has on some transmitted signals.

12-9. (a) Which are the highest and lowest ionic layers above earth? (b) Which layer is most unstable?

12-10. Define the terms *critical frequency* and *maximum usable frequency*.

12-11. Define the term *radiation resistance* and explain how the ohmic value is determined.

12-12. Explain briefly how impedance matching between the line and antenna is obtained using *delta match*.

12-13. Briefly explain why the impedance is not absolute zero at the opened center of the dipole.

12-14. Explain the differences between the terms *driven element* and *parasitic element*, and indicate two types of parasitic units.

12-15. What should be the spacing in inches between the reflector and driven element, as well as between the director and the antenna, for a three-element Yagi? The operating frequency is 50 MHz.

12-16. Give one advantage and one disadvantage of stacking two antenna rods.

12-17. If a stacked pair of antenna rods operates at 60 MHz, what should be the spacing between them (in meters)?

12-18. What methods are employed in antenna stacking to maintain an in-phase condition between elements?

12-19. Briefly explain the difference between a *broadside* array and an *end-fire* array.

12-20. In microwave communications, what advantages are there to using a parabolic reflector?

12-21. How can the energy within a waveguide be propagated from an open end without using an antenna element?

12-22. (a) Are corner reflectors driven elements or parasitic elements? (b) What is the purpose for using a corner reflector?

12-23. In which type of antenna reflector is the antenna placement at the focal point essential? Briefly explain your answer.

12-24. Briefly define the terms *effective radiated power* and *microvolts per meter*.

12-25. What is the field *gain* in decibels for a change from 200 μV/m to 800 μV?

12-26. If the effective radiated power is increased from 1000 to 2000 W, what is the decibel difference?

appendix A
synopsis
of special systems

telemetry. In modern electronic practices, telemetering embraces a number of special modulation systems for information transmission. Basically, the gathering of data is done by measuring and sensing transducers and may include pressure-sensitive devices, speed indicators, temperature-reading units, etc. Such physical changes or analog functions are converted into a representative quantity used for modulating purposes in a process that is more advantageous and convenient than by conventional modulation systems.

Telemetry utilizes multiplexing and permits the transmission of various data (including voice transmissions) in industrial-control applications, space explorations, and other such services. Reference should be made to Chapter 7 for representative multiplexing practices, particularly Sections 7-8, 7-9, and 7-14. Pulse decoding systems, such as described in Section 7-13, should also be reviewed, as well as Section 8-13. Systems descriptions follow.

frequency-division-multiplexing (FDM). This system uses several frequency bands to transmit two or more modulating signals, as described in Sections 7-8 and 7-9. In many instances the separate subcarriers modulated by the signals in turn modulate the primary carrier.

The frequency-division-multiplexing system has been extensively used for increasing the capacity of cable or radio-link devices in international-telephone applications. Some problems have existed with this system, however, particularly in terms of intermodulation, undesired noise signals, and multiple frequency-related filters. Thus, other systems offer many advantages by reducing the undesired factors inherent to the FDM systems for telemetry purposes. Such systems, described next, utilize pulses instead of sine-wave signals for information-conveying purposes. Of considerable aid in the development of such systems has been the solid-state switching circuitry in integrated form to provide for compactness and reliability.

time-division multiplexing (TDM). In this system two or more signals are transmitted by allocating each a finite time interval. Instantaneous amplitudes of the various signals are sampled and transmitted in a time sequence and repeated after the last signal has been sampled. Pulse modulation is used and can consist of any of the processes described next. If the sampling rate is high (such as 8000/s), only negligible signal information is lost, even for voice waveforms, and each channel transmission may be decoded to provide the original signals used for modulation.

pulse-amplitude modulation (PAM). In this system the amplitude of a series of pulses is varied in accordance with the modulating signal, as shown in Fig. A-1, where one of two methods is illustrated. Note that the maximum pulse height is reached for the positive peak of the modulating signal, while the lowest pulse amplitude is at the negative peak of the modulating signal. The other pulse-amplitude modulation method is bidirectional in polarity, instead of unidirectional as in Fig. A-1. Thus, this second system

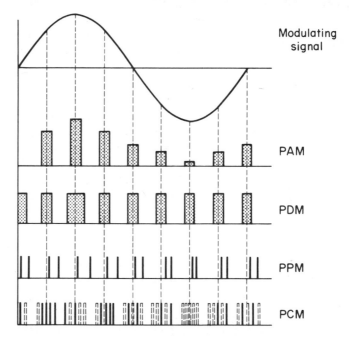

figure A-1

uses pulses having both positive and negative polarities. Thus, the pulse amplitudes for the second alternation of the modulating signal would be identical to the first, except the second set would have a negative polarity.

pulse-duration modulation (PDM). This pulse-modulation system is sometimes referred to as PWM for pulse-width modulation, since the width of progressive pulses is altered to conform to the amplitude of the modulating signal, as also shown in Fig. A-1. Since the pulses have a constant duration, any noise signals riding on the transmitted wavefront in the form of amplitude modulation can be removed by clippers at the receiver. As with other pulse-modulated systems, with narrow-pulse use, additional pulse transmissions can be inserted between the existing ones for increasing the number of information channels. The pulses shown in Fig. A-1 are relatively wider with respect to time for emphasizing the differences between modulating systems.

For the PDM system, the modulation can vary the position of either the leading edge of the pulse or the trailing edge. When such a train of pulses is applied to an integrator circuit, the signal output represents the demodulated waveform. Refer to Fig. 7-16, which shows the process of detection applied to a pulse train containing width modulation. Also note the process illustrated in Fig. 8-19, which utilizes a similar demodulation process.

pulse-position modulation (PPM). This is a form of pulse-time modulation (PTM), since changes in the amplitude of the modulating signal cause a change of the pulse's position in time, as shown in Fig. A-1. The time change can relate to the time between pulses or to the time a pulse occupies in relation to a fixed marker pulse. For Fig. A-1, the marker pulses are those adjacent to the broken vertical lines and are equidistant.

pulse-code modulation (PCM). This system is shown in the lower portion of Fig. A-1, where amplitude variations of the modulating signal are converted to a set of pulses that, in binary form, represent a specific amplitude. This modulation necessitates several sequential processes, one of which is to sample the modulating signals at a predetermined rate. Since the modulating signal contains amplitude variations, the result is a train of pulses also having variations in height.

The amplitude of each pulse so obtained is compared to a selected scale of discrete levels and assigned a value corresponding to the closest found in the comparison process. This procedure is termed *quantizing*, and the values so obtained are translated into a pulse code. The code usually employed is the *binary*, base 2, used extensively in digital computers. Thus, the PCM signals can be processed directly by digital computers. This feature, plus the minimum influence of noise and attenuation on this pulse-modulation method, has made this a practical and widely used system.

In PCM the number of amplitude levels that can be sampled for a single alternation of the modulating signal depends on the number of pulses used in the code group. With a maximum of three pulses, only 8 amplitude variations (from zero) can be referenced, since the binary 111 has a base-ten value of 7. For a code group of four pulses, 16 levels can be sampled (from 0 to 15 inclusive). Table A-1 gives a partial listing for a five-pulse binary code.

table A-1

binary number	base-ten equivalent
00000	0
00001	1
00010	2
00011	3
00100	4
00101	5
00110	6
00111	7
01000	8
01001	9
01010	10
01011	11
01100	12
01101	13
01110	14
01111	15
10000	16
10001	17
10010	18
10011	19
10100	20

The base-ten value given can also be considered as representative of the maximum number of amplitude variations that are possible for a given group of pulses. Note the structure of the binary code: the right-hand vertical number sequence consists of alternate 0's and 1's, the second vertical number sequence has two 0's alternating with two 1's. The third group alternates groups of four 0's and 1's. The coding value from right to left is 1, 2, 4, 8, 16, 32, 64, etc.

appendix B

Standard Broadcast-Station Allocations
AM Band
550 to 1600 KHz
(Nominally 10-kHz bandwidth per station)
FM band
88 to 108 MHz
(200-kHz bandwidth per station)

appendix C
VHF television station frequencies

channel number	frequency in MHz	video carrier	sound carrier
1		not used	
2	54–60	55.25	59.75
3	60–66	61.25	65.75
4	66–72	67.25	71.75
5	76–82	77.25	81.75
6	82–88	83.25	87.75
	(standard FM band 88–108)		
7	174–180	175.25	179.75
8	180–186	181.25	185.75
9	186–192	187.25	191.75
10	192–198	193.25	197.75
11	198–204	199.25	203.75
12	204–210	205.25	209.75
13	210–216	211.25	215.75

appendix D
UHF television
station allocations

channel number	freq. span (MHz)	pix carrier (MHz)	sound carrier (MHz)
14	470–476	471.25	475.75
15	476–482	477.25	481.75
16	482–488	483.25	487.75
17	488–494	489.25	493.75
18	494–500	495.25	499.75
19	500–506	501.25	505.75
20	506–512	507.25	511.75
21	512–518	513.25	517.75
22	518–524	519.25	523.75
23	524–530	525.25	529.75
24	530–536	531.25	535.75
25	536–542	537.25	541.75
26	542–548	543.25	547.75
27	548–554	549.25	553.75
28	554–560	555.25	559.75
29	560–566	561.25	565.75
30	566–572	567.25	571.75
31	572–578	573.25	577.75
32	578–584	579.25	583.75
33	584–590	585.25	589.75
34	590–596	591.25	595.75

channel number	freq. span (MHz)	pix carrier (MHz)	sound carrier (MHz)
35	596–602	597.25	601.75
36	602–608	603.25	607.75
37	608–614	609.25	613.75
38	614–620	615.25	619.75
39	620–626	621.25	625.75
40	626–632	627.25	631.75
41	632–638	633.25	637.75
42	638–644	639.25	643.75
43	644–650	645.25	649.75
44	650–656	651.25	655.75
45	656–662	657.25	661.75
46	662–668	663.25	667.75
47	668–674	669.25	673.75
48	674–680	675.25	679.75
49	680–686	681.25	685.75
50	686–692	687.25	691.75
51	692–698	693.25	697.75
52	698–704	699.25	703.75
53	704–710	705.25	709.75
54	710–716	711.25	715.75
55	716–722	717.25	721.75
56	722–728	723.25	727.75
57	728–734	729.25	733.75
58	734–740	735.25	739.75
59	740–746	741.25	745.75
60	746–752	747.25	751.75
61	752–758	753.25	757.75
62	758–764	759.25	763.75
63	764–770	765.25	769.75
64	770–776	771.25	775.75
65	776–782	777.25	781.75
66	782–788	783.25	787.75
67	788–794	789.25	793.75
68	794–800	795.25	799.75
69	800–806	801.25	805.75
70	806–812	807.25	811.75
71	812–818	813.25	817.75
72	818–824	819.25	823.75
73	824–830	825.25	829.75
74	830–836	831.25	835.75

channel number	freq. span (MHz)	pix carrier (MHz)	sound carrier (MHz)
75	836–842	837.25	841.75
76	842–848	843.25	847.75
77	848–854	849.25	853.75
78	854–860	855.25	859.75
79	860–866	861.25	865.75
80	866–872	867.25	871.75
81	872–878	873.25	877.75
82	878–884	879.25	883.75
83	884–890	885.25	889.75

appendix **E**

television transmission characteristics

television transmission characteristics

One *frame* = 33,334 μs

One field = 16,667 μs

One horixontal sweep cycle (start of one horizontal line trace to start of next) = 63.5 μs

Horizontal blanking interval = 10.16 to 11.4 μs

Horizontal trace (without blanking time) = 53.34 μs

Horizontal sync pulse duration = 5.08 to 5.68 μs

Vertical sync pulse interval (total of six vertical blocks) = 190.5 μs

Vertical blanking interval = 833 to 1300 μs for each field

Vertical scan frequency (monochrome) = 60 Hz

Vertical scan frequency (color) = 59.94 Hz

Horizontal scan frequency (monochrome) = 15,750 Hz

Horizontal scan frequency (color) = 15,734.264 Hz

Total frequency span for an individual station (monochrome or color) = 6 MHz

Picture carrier is nominally 1.25 MHz above the lower end of the channel (monochrome or color)

Vestigial sideband transmission is used for both monochrome and color

Aspect ratio of monochrome or color picture is 4:3

Scan lines per frame (mono or color) = 525 (interlaced)

Scan lines per field (mono or color) = 262.5

The frequency-modulated sound carrier is 4.5 MHz above the picture-carrier frequency (mono or color), with maximum deviation 25 kHz each side of center frequency

The effective radiated power of the audio may range from 50 to 70 per cent of the peak power of the picture-signal carrier

appendix F
color TV factors and terms

The color signal video corresponds to a brightness component transmitted as an AM video carrier, plus a pair of color signals transmitted as AM sidebands procured from a phase-modulated carrier. Hence, two subcarriers 90° apart were involved, having a common frequency related to the color-picture carrier frequency of 3.579,545 MHz.

The modulated color carriers are suppressed at the transmitter, and only sideband signals telecast. An independent carrier oscillator is used at the receiver and synchronized by a color burst signal on the horizontal pedestal. A minimum of eight cycles comprises the burst signal, the frequency of which is identical to the one used at the transmitter: 3.579,545 MHz. The burst signal is omitted following equalizing pulses and is also absent during the vertical pulse interval (or during monochrome transmission).

The brightness, or luminance, portion of the video signal is also referred to as the Y signal.

Hue is the term that identifies a color (such as blue hue, green hue, etc.).

Brightness refers to the intensity of a color.

Saturation refers to the degree to which a color is diluted with white. The less white a color contains, the higher its saturation.

The Y signal consists of predetermined levels of each of the primary colors of red, blue, and green, to correspond to the proportions to which the average eye responds. The Y signal thus contains 0.59 green, 0.30 red, and 0.11 blue.

The I signal consists of combinations of $B - Y$ and $R - Y$ signals, and the proportions are -0.27 blue $- Y$ and 0.74 red $- Y$.

The Q signal (also $B - Y$ and $R - Y$) consists of 0.41 blue $- Y$, and 0.48 red $- Y$.

appendix G
table of trigonometric ratios

degrees	$\dfrac{(opp)}{(hyp)}$ sine	$\dfrac{(adi)}{(hyp)}$ cosine	$\dfrac{(opp)}{(adj)}$ tangent	degrees	$\dfrac{(opp)}{(hyp)}$ sine	$\dfrac{(adj)}{(hyp)}$ cosine	$\dfrac{(opp)}{(adj)}$ tangent
0	0.0000	1.0000	0.0000	20	0.3420	0.9397	0.3640
1	0.0175	0.9998	0.0175	21	0.3584	0.9336	0.3839
2	0.0349	0.9994	0.0349	22	0.3746	0.9272	0.4040
3	0.0523	0.9986	0.0524	23	0.3907	0.9205	0.4245
4	0.0698	0.9976	0.0699	24	0.4067	0.9135	0.4452
5	0.0872	0.9962	0.0875	25	0.4226	0.9063	0.4663
6	0.1045	0.9945	0.1051	26	0.4384	0.8988	0.4877
7	0.1219	0.9925	0.1228	27	0.4540	0.8910	0.5095
8	0.1392	0.9903	0.1405	28	0.4695	0.8829	0.5317
9	0.1564	0.9877	0.1584	29	0.4848	0.8746	0.5543
10	0.1736	0.9848	0.1763	30	0.5000	0.8660	0.5774
11	0.1908	0.9816	0.1944	31	0.5150	0.8572	0.6009
12	0.2079	0.9781	0.2126	32	0.5299	0.8480	0.6249
13	0.2250	0.9744	0.2309	33	0.5446	0.8387	0.6494
14	0.2419	0.9703	0.2493	34	0.5592	0.8290	0.6745
15	0.2588	0.9659	0.2679	35	0.5736	0.8192	0.7002
16	0.2756	0.9613	0.2667	36	0.5878	0.8090	0.7265
17	0.2924	0.9563	0.3057	37	0.6018	0.7986	0.7536
18	0.3090	0.9511	0.3249	38	0.6157	0.7880	0.7813
19	0.3256	0.9455	0.3443	39	0.6293	0.7771	0.8098

degrees	$\frac{(opp)}{(hyp)}$ sine	$\frac{(adj)}{(nyp)}$ cosine	$\frac{(opp)}{(adj)}$ tangent	degrees	$\frac{(opp)}{(hyp)}$ sine	$\frac{(adj)}{(hyp)}$ cosine	$\frac{(opp)}{(adj)}$ tangent
40	0.6428	0.7660	0.8391	65	0.9063	0.4226	2.1445
41	0.6561	0.7547	0.8693	66	0.9135	0.4067	2.2460
42	0.6691	0.7431	0.9004	67	0.9205	0.3907	2.3559
43	0.6820	0.7314	0.9325	68	0.9272	0.3746	2.4751
44	0.6947	0.7193	0.9657	69	0.9336	0.3584	2.6051
45	0.7071	0.7071	1.0000	70	0.9397	0.3420	2.7475
46	0.7193	0.6947	1.0355	71	0.9455	0.3256	2.9042
47	0.7314	0.6820	1.0724	72	0.9511	0.3090	3.0777
48	0.7431	0.6691	1.1106	73	0.9563	0.2924	3.2709
49	0.7547	0.6561	1.1504	74	0.9613	0.2756	3.4874
50	0.7660	0.6428	1.1918	75	0.9659	0.2588	3.7321
51	0.7771	0.6293	1.2349	76	0.9703	0.2419	4.0108
52	0.7880	0.6157	1.2799	77	0.9744	0.2250	4.3315
53	0.7986	0.6018	1.3270	78	0.9781	0.2079	4.7046
54	0.8090	0.5878	1.3764	79	0.9816	0.1908	5.1446
55	0.8192	0.5736	1.4281	80	0.9848	0.1736	5.6713
56	0.8290	0.5592	1.4826	81	0.9877	0.1564	6.3138
57	0.8387	0.5446	1.5399	82	0.9903	0.1392	7.1154
58	0.8480	0.5299	1.6003	83	0.9925	0.1219	8.1443
59	0.8572	0.5150	1.6643	84	0.9945	0.1045	9.5144
60	0.8660	0.5000	1.7321	85	0.9962	0.0872	11.4301
61	0.8746	0.4848	1.8040	86	0.9976	0.0698	14.3007
62	0.8829	0.4695	1.8807	87	0.9986	0.0523	19.0811
63	0.8910	0.4540	1.9626	88	0.9994	0.0349	28.6363
64	0.8988	0.4384	2.0503	89	0.9998	0.0175	57.2900
65	0.9063	0.4226	2.1445	90	1.0000	0.0000	. . .

appendix **H**

table of common logarithm

N	0	1	2	3	4	5	6	7	8	9
10	0000	0043	0086	0128*	0170	0212	0253	0294	0334	0374
11	0414	0453	0492	0531	0569	0607	0645	0682	0719	0755
12	0972	0828	0864	0899	0934	0969	1004	1038	1072	1106
13	1139	1173	1206	1239	1271	1303	1335	1367	1399	1430
14	1461	1492	1523	1553	1584	1614	1644	1673	1703	1732
15	1761	1790	1818	1847	1875	1903	1931	1959	1987	2014
16	2041	2068	2095	2122	2148	2175	2201	2227	2253	2279
17	2304	2330	2355	2380	2405	2430	2455	2480	2504	2529
18	2553	2577	2601	2625	2648	2672	2695	2718	2742	2765
19	2788	2810	2833	2856	2878	2900	2923	2945	2967	2989
20	3010	3032	3054	3075	3096	3118	3139	3160	3181	3201
21	3222	3243	3263	3284	3304	3324	3345	3365	3385	3404
22	3423	3444	3464	3483	3502	3522	3541	3560	3579	3598
23	3617	3636	3655	3674	3692	3711	3729	3747	3766	3784
24	3802	3820	3838	3856	3874	3892	3909	3927	3945	3962
25	3979	3997	4014	4031	4048	4065	4082	4099	4116	4133
26	4150	4166	4183	4200	4216	4232	4249	4265	4281	4298
27	4314	4330	4346	4362	4378	4393	4409	4425	4440	4456
28	4472	4487	4502	4518	4533	4548	4564	4579	4594	4609
29	4624	4639	4654	4669	4683	4698	4713	4728	4742	4757
N	0	1	2	3	4	5	6	7	8	9

N	0	1	2	3	4	5	6	7	8	9
30	4771	4786	4800	4814	4829	4843	4857	4871	4886	4900
31	4914	4928	4942	4955	4969	4983	4997	5011	5024	5038
32	5051	5065	5079	5092	5105	5119	5132	5145	5159	5172
33	5185	5198	5211	5224	5237	5250	5263	5276	5289	5302
34	5315	5328	5340	5353	5366	5378	5391	5403	5416	5428
35	5441	5453	5465	5478	5490	5502	5514	5527	5539	5551
36	5563	5575	5587	5599	5611	5623	5635	5647	5658	5670
37	5682	5694	5705	5717	5729	5740	5752	5763	5775	5786
38	5798	5809	5821	5832	5843	5855	5866	5877	5888	5899
39	5911	5922	5933	5944	5955	5966	5977	5988	5999	6010
40	6021	6031	6042	6053	6064	6075	6085	6096	6107	6117
41	6128	6138	6149	6160	6170	6180	6191	6201	6212	6222
42	6232	6243	6253	6263	6274	6284	6294	6304	6314	6325
43	6335	6345	6355	6365	6375	6385	6395	6405	6415	6425
44	6435	6444	6454	6464	6474	6484	6493	6503	6513	6522
45	6532	6542	6551	6561	6571	6580	6590	6599	6609	6618
46	6628	6637	6646	6656	6665	6675	6684	6693	6702	6712
47	6721	6730	6739	6749	6758	6767	6776	6785	6794	6803
48	6812	6821	6830	6839	6848	6857	6866	6875	6884	6893
49	6902	6911	6920	6928	6937	6946	6955	6964	6972	6981
50	6990	6998	7007	7016	7024	7033	7042	7050	7059	7067
51	7076	7084	7093	7101	7110	7118	7126	7135	7143	7152
52	7160	7168	7177	7185	7193	7202	7210	7218	7226	7235
53	7243	7251	7259	7267	7275	7284	7292	7300	7308	7316
54	7324	7332	7340	7348	7356	7364	7372	7380	7388	7396
55	7404	7412	7419	7427	7435	7443	7451	7459	7466	7474
56	7482	7490	7497	7505	7513	7520	7528	7536	7543	7551
57	7559	7566	7574	7582	7589	7597	7604	7612	7619	7627
58	7634	7642	7649	7657	7664	7672	7679	7686	7694	7701
59	7709	7716	7723	7731	7738	7745	7752	7760	7767	7774
60	7782	7789	7796	7803	7810	7818	7825	7832	7839	7846
61	7853	7860	7868	7875	7882	7889	7896	7903	7910	7917
62	7924	7931	7938	7945	7951	7959	7966	7973	7980	7987
63	7993	8000	8007	8014	8021	8028	8035	8041	8048	8055
64	8062	8069	8075	8082	8089	8096	8102	8109	8116	8122
N	0	1	2	3	4	5	6	7	8	9

N	0	1	2	3	4	5	6	7	8	9
65	8129	8136	8142	8149	8156	8162	8169	8176	8182	8189
66	8195	8202	8209	8215	8222	8228	8235	8241	8248	8254
67	8261	8267	8274	8280	8287	8293	8299	8306	8312	8319
68	8325	8331	8338	8344	8351	8357	8363	8370	8376	8382
69	8388	8395	8401	8407	8414	8420	8426	8432	8439	8445
70	8451	8457	8463	8470	8476	8482	8488	8494	8500	8506
71	8513	8519	8525	8531	8537	8543	8549	8555	8561	8567
72	9573	9579	8585	8591	8597	8603	8609	8615	8621	8627
73	8633	8639	8645	8651	8657	8663	8669	9675	8681	8686
74	8692	8692	8698	8704	8710	8716	8722	8733	8739	8745
75	8751	8756	8762	8768	8774	8779	8785	8791	8797	8802
76	8808	8814	8820	8825	8831	8837	8842	8848	8854	8859
77	8865	8871	8876	8882	8887	8893	8899	8904	8910	8915
78	8921	8927	8932	8938	8943	8949	8954	8960	8965	8971
79	8976	8982	8987	8993	8998	9004	9009	9015	9020	9025
80	9031	9036	9042	9047	9053	9058	9063	9069	9074	9079
81	9085	9090	9096	9101	9106	9112	9117	9122	9128	9133
82	9138	9143	9149	9154	9159	9165	9170	9175	9180	9186
83	9191	9196	9201	9206	9212	9217	9222	9227	9232	9238
84	9243	9248	9253	9258	9263	9269	9274	9279	9284	9289
85	9294	9299	9304	9309	9315	9320	9325	9330	9335	9340
86	9345	9350	9355	9360	9365	9370	9375	9380	9385	9390
87	9395	9400	9405	9410	9415	9420	9425	9430	9435	9440
88	9445	9450	9455	9460	9465	9469	9474	9479	9484	9489
89	4949	9499	9504	9509	9513	9518	9523	9528	9533	9538
90	9542	9547	9552	9557	9562	9566	9571	9576	9581	9586
91	9590	9595	9600	9605	9609	9614	9619	9624	9628	9633
92	9638	9643	9647	9652	9657	9661	9666	9671	9675	9680
93	9685	9689	9694	9699	9703	9708	9713	9717	9722	9727
94	9731	9736	9741	9745	9750	9754	9759	9763	9768	9773
95	9777	9782	9786	9791	9795	9800	9805	9809	9814	9818
96	9823	9827	9832	9836	9841	9845	9850	9854	9859	9863
97	9868	9872	9877	9881	9886	9890	9894	9899	9903	9908
98	9912	9917	9921	9921	9926	9930	9939	9943	9948	9952
99	9956	9961	9965	9969	9974	9978	9983	9987	9991	9996
N	0	1	2	3	4	5	6	7	8	9

appendix I
exponential function

x	e^x	e^{-x}	$\sinh$ x	$\cosh$ x	$\tanh$ x	$\coth$ x	$\sinh^{-1}$ x	$\cosh^{-1}$ x	$\tanh^{-1}$ x	$\coth^{-1}$ x
0.00	1.000	1.000	0.000	1.000	0.000	∞	0.000	. . .	0.000	
0.10	1.105	0.905	0.100	1.005	0.100	10.033	0.100	. . .	0.100	
0.20	1.221	0.819	0.201	1.020	0.197	5.066	0.199	. . .	0.203	
0.30	1.350	0.741	0.305	1.045	0.291	3.433	0.296	. . .	0.309	
0.40	1.492	0.670	0.411	1.081	0.380	2.632	0.390	. . .	0.424	
0.50	1.649	0.607	0.521	1.128	0.462	2.164	0.481	. . .	0.549	
0.60	1.822	0.549	0.637	1.185	0.537	1.862	0.569	. . .	0.693	
0.70	2.014	0.497	0.759	1.255	0.604	1.655	0.653	. . .	0.867	
0.80	2.226	0.449	0.888	1.337	0.664	1.506	0.733	. . .	1.099	
0.90	2.460	0.407	1.027	1.433	0.716	1.396	0.809	. . .	1.472	
1.00	2.718	0.368	1.175	1.543	0.762	1.313	0.881	0.000	∞	∞
1.10	3.004	0.333	1.336	1.669	0.800	1.249	0.950	0.444	. . .	1.522
1.20	3.320	0.301	1.509	1.811	0.834	1.200	1.016	0.622	. . .	1.199
1.30	3.669	0.273	1.698	1.971	0.862	1.160	1.079	0.756	. . .	1.018
1.40	4.055	0.247	1.904	2.151	0.885	1.129	1.138	0.867	. . .	0.896
1.50	4.482	0.223	2.129	2.352	0.905	1.105	1.195	0.962	. . .	0.805
1.60	4.953	0.202	2.376	2.577	0.922	1.085	1.249	1.047	. . .	0.733
1.70	5.474	0.183	2.646	2.828	0.935	1.069	1.301	1.123	. . .	0.675
1.80	6.050	0.165	2.942	3.107	0.947	1.056	1.350	1.193	. . .	0.626
1.90	6.686	0.150	3.268	3.418	0.956	1.046	1.398	1.257	. . .	0.585
2.00	7.389	0.135	3.627	3.762	0.964	1.037	1.444	1.317	. . .	0.549

x	e^x	e^{-x}	sinh x	cosh x	tanh x	coth x	sinh^{-1} x	cosh^{-1} x	tanh^{-1} x	coth^{-1} x
2.10	8.166	0.122	4.022	4.144	0.970	1.030	1.487	1.373	...	0.518
2.20	9.025	0.111	4.457	4.568	0.976	1.025	1.530	1.425	...	0.490
2.30	9.974	0.100	4.937	5.037	0.980	1.020	1.570	1.475	...	0.466
2.40	11.02	0.091	5.466	5.557	0.984	1.017	1.609	1.522	...	0.444
2.50	12.18	0.082	6.050	6.132	0.987	1.014	1.647	1.567	...	0.424
2.60	13.46	0.074	6.695	6.769	0.989	1.011	1.684	1.609	...	0.405
2.70	14.88	0.067	7.406	7.473	0.991	1.009	1.719	1.650	...	0.389
2.80	16.44	0.061	8.192	8.253	0.993	1.007	1.753	1.689	...	0.374
2.90	18.17	0.055	9.060	9.115	0.994	1.006	1.786	1.727	...	0.360
3.00	20.09	0.050	10.018	10.068	0.995	1.005	1.818	1.763	...	0.347
3.10	22.20	0.045	11.08	11.12	0.996	1.004	1.850	1.798	...	0.335
3.20	24.53	0.041	12.25	12.29	0.997	1.003	1.880	1.831	...	0.323
3.30	27.11	0.037	13.54	13.57	0.997	1.003	1.909	1.863	...	0.313
3.40	29.96	0.033	14.97	15.00	0.998	1.002	1.938	1.895	...	0.303
3.50	33.12	0.030	16.54	16.57	0.998	1.002	1.966	1.925	...	0.294
4.00	54.60	0.018	27.29	27.31	0.999	1.001	2.095	2.063	...	0.255
4.50	90.02	0.0111	45.00	45.01	1.000	1.000	2.209	2.185	...	0.226
5.00	148.4	0.0067	74.20	74.21	1.000	1.000	2.312	2.292	...	0.203
5.50	244.7	0.0041	122.3	122.3	1.000	1.000	2.406	2.390	...	0.184
6.00	403.4	0.0025	201.7	201.7	1.000	1.000	2.492	2.478	...	0.168

appendix J
powers-of-ten designations

atto		$1 - 10^{-18}$
femto		$1 - 10^{-15}$
pico (micro-micro)	one-millionth of a million	$1 - 10^{-12}$
nano (milli-micro)	one-thousandth of a million	$1 - 10^{-9}$
micro	one-millionth	$1 - 10^{-6}$
milli	one-thousandth	$1 - 10^{-3}$
centi	one-hundredth	$1 - 10^{-2}$
deci	one-tenth	$1 - 10^{-1}$
deca	ten	$1 - 10^{1}$
hecto	one hundred	$1 - 10^{2}$
kilo	one thousand	$1 - 10^{3}$
mega	one million	$1 - 10^{6}$
giga	one thousand million	$1 - 10^{9}$
tera	one million million	$1 - 10^{12}$

index